AUSTRALIAN MAMMALS
IN COLOUR

AUSTRALIAN MAMMALS

IN COLOUR

Irene & Michael Morcombe

REED

First published 1979

A. H. & A. W. REED PTY LTD
53 Myoora Road, Terrey Hills, Sydney
65-67 Taranaki Street, Wellington
11 Southampton Row, London
also at
Auckland and Christchurch

National Library of Australia
Cataloguing in Publication data
Morcombe, Irene
Australian mammals in colour.
(Reeds colourbook series)
Index
ISBN 0 589 50084 8
1. Mammals—Australia—Pictorial works.
I. Morcombe, Michael K., joint author.
II. Title. (Series).
559'.0994

Set by ASA Typesetters, Sydney
Printed and bound by
Kyodo-Shing Loong Printing Industries Pte Ltd, Singapore

Contents

Introduction

AUSTRALIA together with adjacent New Guinea is the only place on earth where the three basic groups of mammals still exist side by side. In order of supposed evolutionary status, these groups are the monotremes which are considered comparatively primitive kinds of mammals, the marsupials, and the placentals.

Of these groups, only the monotremes are unique to the Australian region. The marsupials, while being by far most numerous and diverse in Australia, are also represented by a few species in both North and South America. The placental mammals on the other hand, while being represented in Australia, are a small minority here while being the dominant or only kind of mammal on most other continents.

The existence and relative importance of each of these mammal groups in Australia can be explained only by looking far back into the geological history of this part of the earth. South America, although separated by the width of the Pacific, has marsupials, but Asia close to the north has not. South Africa, separated by the width of the Indian Ocean, shares with Australia many similarities of plant life, especially in the family Proteaceae. The genus *Nothofagus*, of which the Antarctic Beech is an example, is found in Australia, New Guinea, New Zealand, South America, and through fossil evidence is known to have occurred in Antarctica.

The answer lies in the now generally accepted fact of continental drift. Australia, New Guinea, Africa, India, South America and Antarctica were linked as a single very large continent about 150 million years ago, having much earlier separated from the northern continents. Although the movement of Australia and New Guinea together northwards has been only at some 25 millimetres a year, this amounts to some 19,000 kilometres in a hundred million years, not a long time in geological terms. The marsupials first appeared while Australia and South America were still joined, hence the presence of marsupials on both.

The similarities in flora also date back to times when the southern continents were one.

Australia and New Guinea together drifted northwards for some 30 million years in complete isolation and surrounded by very wide ocean barriers. During this time their animal life was cut off from outside influences and competition. The marsupials evolved towards their present great diversity of species, producing marsupial carnivores, herbivores and specialist inhabitants of tree-tops, plains, rocky ranges and deserts.

Many of these animals of Australia and New Guinea closely paralleled equivalent types of placental mammals at the same time evolving on the northern continents, so that there were marsupial cats, marsupial moles, marsupial herbivores, gliders and possums, with equivalent feline cats, ordinary moles, herbivores, gliding squirrels, and monkeys occupying the same environmental niches elsewhere.

Eventually, some 10 million years ago, the closing gap between New Guinea and Asian islands allowed a succession of Asian placental mammals to find their way into Australia. Some may have come by accidental crossing of the sea gaps as these became narrower—windblown bats, flying foxes, and of course birds. Small animals may have crossed on drifting debris after storms, and it is interesting to note that these placental mammals native to Australia are small. Lowered sea levels during a succession of ice ages probably bridged some gaps such as Torres Strait, others were too deep and remained a permanent barrier to the larger Asian mammals.

Early mammal invaders have now been in Australia so long that they have become unique to this continent, though of course they are recent by comparison with marsupials and monotremes. Later arrivals still show a close affinity to mammals of the rest of the world, although new species have evolved here.

One of the more recent arrivals from the north was Aboriginal man, a placental mammal like the preceding succession of invaders. He also adapted to the Australian environment, became integrated into the overall fauna as a new predator, but struck a balance with the land and its creatures that could have been maintained almost indefinitely without damage to the environment.

Perhaps a greater effect was caused by a species of dog which early man is thought to have brought with him as a domesticated animal. Being the only large non-marsupial carnivore in Australia, this dog, the Dingo, probably competed with the marsupial carnivores, and perhaps together with gradual climatic changes, may have been partly responsible for the disappearance of the Tasmanian Tiger and Devil from the Australian mainland.

The rate of invasion by placental mammals has accelerated enormously following the arrival of European man. The effect, to the detriment of the native mammals, has been greater than ever previously encountered.

The three groups of mammals which Australia and New Guinea still preserve are separated principally by differences of reproductive systems. Although all three originally evolved from reptile-like ancestors, the monotremes are considered most primitive in their similarities to the reptiles. The bones of their shoulder region are arranged like the shoulder girdle of the reptiles, and the female monotreme lays eggs like those of the reptiles. The name 'monotreme' is given because there is but one ventral opening for the two functions of reproduction and elimination.

Placental and marsupial kinds of mammals both give birth to living young rather than lay eggs, but while the young of the placental mammal are large and advanced in development (in some species able to run with the mother within hours of birth) those of the marsupial are tiny and born at an embryonic stage in development; the marsupial pouch is used to complete development to the stage similar to that at which the placental young is born.

Australia's monotremes and marsupials, so different from the animals of other continents, always have fascinated both the early explorers and later visitors to the region. Stories reaching Europe of egg-laying, duck-billed furred animals were not believed, skins were regarded as fakes. Only after a long time was it realised that the Platypus and Spiny Anteater were not weird mixtures of duck, fish, mammal or whatever, but were on the one hand extremely primitive types of mammal, and on the other hand ones which had become very highly specialised for a particular way of life leading to development of beak-like snouts, webbed feet and so on.

In the case of marsupials, the pouch has always been a fascinating feature, so that the term 'pouched animals' is almost synonymous with 'marsupial', although with some species the mammary area does not form the typical deep pouch such as that possessed by kangaroos. Because the young in the pouch are at first only a few millimetres in length when first noticed it was thought that they were actually born in the pouch, coming into being as tiny bud-like swellings at the tips of the teats.

It has long been recognised that this is anatomically impossible, but the truth is more astounding than the fiction. The young marsupial, incredibly small compared with its parent, embryonic and disproportionate in shape, unable to see or hear, after being born in the normal manner struggles through the fur (being too small to go over the fur) and usually but not always finds its way into the pouch to attach to a teat and there to remain, until as large as a placental mammal is around the time of its birth. The tiny creature does not always make the distance, sometimes being lost in the wrong direction or being knocked to the ground by some careless movement of the female marsupial, which does not assist but generally sits quietly during the time, between thirty minutes and an hour, from birth to entering the pouch.

Although extremely premature, embryonic in fact compared with other animals, the tiny marsupial is well developed in those features necessary to get it to the pouch, particularly the forelimbs, jaws and tongue. While the eyes and ears are not developed, the nostrils are large, suggesting that the sense of smell may assist it to find the way to the pouch.

This feat of the young marsupial making the journey unaided to the pouch is all the more remarkable when its size is considered—the newborn joey of the Bennetts Wallaby, quite a large wallaby, is a mere 18 mm total head and body length.

Since the time of the earliest occupation of Australia by European man, he has almost wherever possible endeavoured to obliterate the harsh Australian environment and replace it with European derived surroundings. Forests have been destroyed, and a great many foreign species of animals and plants introduced. Many of these have become serious pests, and have become so securely established that they have proved almost impossible to eradicate. Many cause serious economic losses, most compete in one way or another with native species. All too often the introduced species, freed of the enemies and diseases of their land of origin, are overwhelmingly successful, and displace the Australian native species.

Herds of cattle and mobs of sheep on semi-arid inland plains have had a disastrous effect, eating out the low ground cover scrub that small native marsupials and rodents need for shelter or food, and have trampled the earth so hard-packed that it is no longer habitable for small burrowing animals.

Marsupials which represent some 50 million years of evolution have gone entirely from the face of the earth, others have become extremely rare. Some thought to have been lost have been rediscovered, but there are others which dwelt where now their bushland habitat has been entirely replaced by grassed paddocks and barbed wire fences and are certainly gone for all time. The future of a great many others of Australia's unique mammals will depend very much upon how we care for the natural environment. It is not the obvious devastation that is always worst, but the insidious steady deterioration caused by over-grazing, overly frequent burning of forests and scrublands. Australia is for the most part a very arid continent, where such mismanagement of the environment is causing not only a loss of the mammals but also a steady trend towards total and useless desert.

Tiger Cat

Dasyurus maculatus

THE largest of the marsupial carnivores now inhabiting mainland Australia, the Tiger Cat, like so many of Australia's native mammals, carries a name which is an inaccurate, yet in some ways an effective, description. It is not a cat, and it is unrelated to the felines that evolved on other continents. Nor does it bear comparison with the tiger, in size or in pattern of coat.

But what other descriptive title can be given this nocturnal hunter which is rather cat-like in superficial appearance, and has a reputation for fearlessness and ferocity that earns the tiger comparison.

The similarity between this Dasyurid carnivore and the feline carnivores comes not through any common ancestry, for one is a marsupial and the others are placental mammals—of entirely different sub-classes of mammals. It is the result, rather, off the forces of convergent evolution. The more primitive marsupials of this isolated Australian island, and later the advanced placental mammals of other land masses, both exploited similar opportunities in their separate parts of the world, following a similar stealthy predatory way of life. As a consequence they have similar proportions, lithe, supple and muscular bodies, similar flesh-tearing teeth, paws sharply-clawed, and an overall likeness.

The marsupial carnivores are not cats, but remarkably cat-like by consequence of their cat-like way of life. For this resemblance and not for any relationship are they called 'native cats'. They are also quite commonly referred to as 'dasyurids', a term meaningful to zoologists but conveying no descriptive message to the vast majority of people unfamiliar with scientific nomenclature of Australian marsupials.

The Tiger Cat is by far the largest of the four species of native cats, and it is the only species with spots on its tail as well as its body. It is generally somewhat larger than the ordinary domestic cat, with a body length often reaching 60 cm and a tail of 50 cm. Compared with the Eastern Native-Cat, the only other species occurring over the major part of the Tiger Cat's range of distribution, it also differs from that smaller carnivore in having five toes on the hind feet instead of four.

The largest of the tree-climbing native carnivores, Tiger Cats are said to be capable of taking quite large prey, even the smaller species of wallabies. They inhabit dense forests, heavy scrubs and rainforests of the coasts and mountain ranges, and occur in north-eastern and south-eastern Queensland, eastern New South Wales, southern Victoria, the south-east of South Australia, and throughout Tasmania.

Tiger Cats are skilful climbers, and are most often seen in trees. They are well equipped for climbing with strong claws and serrated pads on the palms of front and hind feet.

The young, four to six in number, are born, as usual with marsupials, extremely small, and are carried in a kind of shallow pouch, a flap of skin covering the side and front of the mammary area.

Tiger Cats have become uncommon in most regions with the felling of forests and other activities destructive of their habitat. They are now found only in the wildest of rugged heavily forested areas, and are most common in Tasmania.

Tasmanian Devil

Sarcophilus harrisii

SOLID thick-set and powerful, the Devil is, though little more than a metre in length, exceeded in size only by the *Thylacine* or Tasmanian Tiger. Its head, like that of the Hyaena, seems disproportionately large for its body, the muzzle is broad, the jaws wide and powerful, and even the ears appropriately stumpy and rounded.

Add to this that the Devil is black (but for several white bands), utters unpleasant rasping and snarling sounds, and moves in a strange loping run, again somehow hyaena-like, and a picture emerges of an animal which appears as the most unpleasant, ferocious-looking of all Australian native mammals, and its name seems very appropriate indeed.

But the Tasmanian Devil is principally a scavenger. Herein lies the reason that it shares to some degree the build and features of the Hyaena, which is its counterpart among the placental mammals.

The Tasmanian Devil will consume almost any carrion it can find, perhaps in the past the left-overs from the kill of the *Thylacine*. Almost every part of a carcass, the bones (but for small hard bits such as jaws bearing teeth) are crushed and consumed. Reptiles, fish and crustacea will be eaten if the opportunity arises, and although a carrion-eater, the Devil does not hesitate to kill for itself anything that is not too large or too fast.

Living prey includes small mammals, reptiles, and even full-grown sheep if these are penned or weak, and unable to escape. There is a record of a chained-up sheep dog pup being eaten.

Fossil deposits and cave remains indicate that, although now confined to Tasmania, the Tasmanian Devil was common on the Australian mainland until comparatively recent times. There is some conjecture that this species may still occur in southern Victoria. Bones found among remains of present-day marsupials in kitchen middens of Aboriginals show that Tasmanian Devils inhabited that region during inhabitation by Aboriginal man. A specimen was killed at Toobarac in 1912, and although this is generally believed to have been an escapee from a zoo, sanctuary or other captivity, this has never been substantiated. In Tasmania it is an inhabitant of eucalypt forests, usually in rocky areas.

The usual marsupial pouch is present, consisting of a horseshoe shaped flap of skin, the opening of which is towards the rear. When the young are about five months old the female Devil builds a grass bed for them in a hollow log or crevice among boulders.

Although the Tasmanian Devil has commonly been described as savage, when reared in captivity it is said to make a docile, affectionate pet.

Western Native-cat

Dasyurus geoffroii

ALTHOUGH superficially similar to the Tiger Cat, the three *Dasyurus* species known as 'native cats' have a number of differentiating characteristics apart from their much smaller size. Most obvious is the lack of any spots on their tails, although the Eastern Native-cat, or Quoll, may have a white tail-tip.

The native cats and tiger cat are now all species of the genus *Dasyurus*, but previously have been classified under four generic names, the Western Native-cat being *Dasyurinus*, the Tiger Cat *Dasyurops*, the Little Northern Native-cat *Satanellus*, and only the Eastern Native-cat, *Dasyurus quoll*, having the name that is now applied to the whole group.

The Western Native-cat is not only an inhabitant of Western Australia. Its distribution includes the whole of that State except the Kimberley region, all of South Australia, the southern half of the Northern Territory, Queensland except for the far north and north-east, north-western Victoria, and New South Wales west of the Great Dividing Range. Obviously it was called 'Western Native-cat' because it was found to the west of the Great Divide in the early days, but today it is very rare or extinct except in south-western Australia.

Fortunately the native cat remains quite common in the heavily forested south-western corner of Australia, the Jarrah and Karri forests which extend southwards from the vicinity of Perth on the west coast to Albany on the south coast. It still occurs in the Darling Ranges within fifteen or twenty kilometres of Perth. These south-western native cats are of a slightly different sub-species to the Western Native-cats which formerly roamed across the open woodland country of the eastern States of Australia, being slightly larger, and lighter coloured underneath.

Eastwards of the Great Dividing Range, in southern Victoria, the south-east of South Australia, Tasmania and King Island, the native cat tribe is represented by the Eastern Native-cat, *Dasyurus viverrinus*. Across far northern Australia, from the Pilbara and Kimberley region of Western Australia through the Northern Territory to north-eastern Queensland, is found the Little Northern Native-cat, *Dasyurus hallucatus*.

The Little Northern Native-cat, much smaller than the preceding species, is regarded as the most primitive, the most like the basic insectivorous marsupial mice from which Australia's varied marsupial population is thought to have evolved.

The native cats are usually nocturnal, but may be seen occasionally foraging out of their tree hollow or rock crevice lairs in late afternoon or early morning.

Although having a reputation for attacking and killing poultry, the native cats are also beneficial in feeding much more commonly upon mice and grasshoppers. They must also of course take many nesting and roosting native birds, but this is within the framework of the natural balance of nature.

Although as many as eighteen extremely tiny young may be born, there are places in the pouch for eight only, presumably the first-born and strongest. Only those which attach to a teat can survive.

When too large for the pouch the young cling all over their mother, eventually almost dragging her to the ground with their weight as she tries to hunt at night.

Fat-tailed Dunnart

Sminthopsis crassicaudata

THE MARSUPIAL mice are, among the present-day marsupials of Australia, most like the first marsupials inhabiting this continent. They are probably the most primitive and generalised of the marsupials, insect-eaters with many small sharp unspecialised front teeth. Their paws retain the earliest five-digit arrangement, unlike the plant-eating marsupials which have evolved a fusion of second and third toes on the hind feet. The plant-eating marsupials, the kangaroos and possums, have developed a specialised dentition, with a pair of chisel-like incisors standing out from the other teeth.

But although the marsupial mice are the most primitive of marsupials, their appearance marked a big advance over the egg-laying monotremes. Compared with the monotremes (which include the platypus and spiny anteater) they are more warm-blooded and hence more active and agile.

The monotremes, more reptile-like, carry their bodies close to the ground, legs spreading outwards to each side, while the marsupial mice stand poised more directly above their legs. Their external ear flaps give keener perception and direction of sounds than the ear holes of the monotremes and the reptiles. Above all in reproductive efficiency and evolution of brain the marsupial mice represented a major step, a successful model from which evolved the great diversity of other marsupials which have inhabited this continent.

The Dunnart, which can immediately be recognised by its fattened tail, belongs to a group having extremely narrow and elongated hind feet, the genus *Sminthopsis*. These mouse-sized animals are active nocturnal insect-eaters, which at times will tackle larger prey such as ordinary mice, small birds and lizards.

There are probably eleven species of dunnarts with tails which may become fattened at times when food is abundant, but at other times may be captured with quite thin tails. These are for the most part inhabitants of arid or semi-arid country where the tail food store is a valuable survival characteristic in times of drought.

Of all the species the Fat-tailed Dunnart has been most studied. Its range of distribution, inland south-eastern and south-western Australia, is less remote than that of most other species.

Like some other small mammals it is able to lower its body temperature at times of food shortage, becoming sluggish and remaining inactive in its nest, keeping to a minimum its energy expenditure. During daylight hours the Dunnart hides in a small hollow log, under a stone or in a clump of dense vegetation.

The female may start to breed at the age of four months, and, in good seasons, produces litters continuously at intervals of about eighty-two days. These tiny young are carried in a pouch.

Numbat

Myrmecobius fasciatus

THIS MARSUPIAL anteater is one of the most beautifully coloured of Australian mammals, with its rusty-red and rich browns accentuated by bands and stripes of white and black.

There are two slightly differing sub-species, the Numbat of the south-western corner of the continent, *Myrmecobius fasciatus fasciatus*, and the Rusty Numbat, *Myrmecobius fasciatus rufus*, which inhabits the remote north-western parts of South Australia and adjacent border areas of Western Australia.

Together these ranged across southern Australia, from Western Australia to the western parts of New South Wales. But they have long been extinct in eastern Australia, and are now restricted to remnant patches of forest in south-western Australia, and arid ranges of the north-west of South Australia.

Unlike the majority of Australian marsupials, the Numbat is abroad during daylight hours, busily searching for the termites which are its only food. Possums, gliders, native cats and other native marsupials must be searched for by spotlight, at night, but in a few forests of south-western Australia Numbats may be encountered during the day. They are most often seen dashing across bush tracks ahead of a motor vehicle, more rarely are they seen by even a stealthy observer on foot.

When feeding, scratching aside fallen twigs to expose termite trails, or digging into a termite mound, the Numbat is constantly alert for the slightest sound of danger. It frequently pauses, listening with sensitive ears, and will usually adopt a very characteristic pose, with one forepaw lifted above ground. Occasionally, trying to determine the source of some faint sound or movement, it will stand upright upon its two hind legs to gain a little more height. If there is danger, the Numbat will run swiftly to one of the many hollow logs that litter the ground in a typical Numbat territory, or on some occasions will dive into a rabbit burrow.

It appears that young Numbats are born late summer to autumn. The normal litter is four, and these are born extremely tiny, as is usual with marsupials. They cling to the long coarse hairs of their mother's pouch area with their forepaws, and are also held securely by the teats in their mouths, and these they will not release. When they are tiny, the female Numbat pays no attention to them, and seems almost to be unaware of their presence. Later, when they are old enough to leave her, she conceals them in a hollow log while she is out feeding.

The Numbat is related to the native cats, but has evolved special features to suit its specialised way of life, enabling it to fill a niche not exploited by any other Australian mammal. It no longer resembles the other marsupial carnivores in appearance. The divergence is so great that the Numbat was for a time placed in a separate family of its own.

The principal adaptations to the specialised termite diet are the modified teeth and the long slender tongue.

Dibbler

Antechinus apicalis

THE NAME 'Dibbler' was used by Aboriginals who brought specimens of this animal to John Gilbert, collector of zoological specimens in the colony in 1838. About 23 cm in total length, these small carnivorous marsupials had a distinctive white ring of fur around each eye, speckled fur, and a tail that tapered from the base, where the fur was longest.

The Dibbler was first recorded from specimens collected near the Moore River, north of Perth. The natives here said that the 'Dibbler' heaped up a nest of grass and sticks larger than that built by the Quenda. In the following years other collectors obtained Dibblers in the vicinity of Albany, but no further information of the creature's habits, and very little about the precise collecting localities, was recorded. The last specimen on record was one obtained by the National Museum of Victoria in 1884.

After eighty years in which there was no sign that this species still existed, it was thought to be extinct. Quite apart from its rarity, the species was of considerable scientific interest because so little had in the early collecting days been recorded of its biology, and because it appeared to be a connecting link between the small insectivorous marsupial mice, and the larger carnivorous native cats.

The Dibbler was rediscovered when a female was captured by the author of this book, on the night of 26 January 1967, on the lower slopes of Mt Manypeaks, east of Albany. The rediscovery was made accidentally during an attempt to find Honey Possums. This Dibbler, and several others caught soon after, was taken at the flowers of the shrub *Banksia attenuata*, growing in dense low scrub thickets in coastal sand heath country. From 1967 no living specimen of the Dibbler has been captured except at precisely the same locality of this rediscovery, although a number of attempts have been made, and much cage-trapping undertaken in a variety of habitats along the south coast of Western Australia. There has been a single indication that there must be other populations—the bringing in, by a cat, of a dead specimen, many kilometres eastwards of the Mt Manypeaks habitat.

From these discoveries, and subsequent studies of the animals in captivity, much has been added to the early sketchy knowledge of the biology of the species.

Because some were captured on banksia flowers, it is known that they are partly arboreal, and that the large wildflowers are visited for nectar or for the insects which are attracted to the flowers. The Dibbler may not be entirely nocturnal, but may move around under the dense layer of leaf litter under the thick low scrub for some hours after sunrise and before sunset. Observations did not show that it built a nest, like that described to Gilbert by the natives in 1838, but rather that it would push peacefully into dense accumulations of debris, which perhaps could be mistaken for a nest.

Great Grey or Forester

Macropus giganteus

THE GREAT GREY is probably the best known of the big large kangaroos because its distribution coincides with the areas of heaviest human population. It is found almost throughout eastern Australia from Cape York Peninsula southwards, covering the eastern half of Queensland, most of New South Wales except parts of the far west, almost all of Victoria, and the northern and eastern parts of Tasmania.

Within this wide range of distribution the Great Grey not surprisingly shows some variation and there are several sub-species. The Great Grey was first described from a specimen from the north of its range, so that it is the race *Macropus giganteus giganteus* that is found in northern Queensland, the race *M. g. major* the southern part, and the race *M. g. tasmaniensis* that inhabits Tasmania.

The Tasmanian Great Grey or Forester is distinctive in being usually more brown rather than grey, and the fur is longer and more shaggy. A similar longer fur is seen among Great Greys from cold highland areas of south-eastern Australia.

The Great Grey is a species quite separate from the Western Grey Kangaroo, which occurs not only in Western Australia, but also in parts of south-eastern Australia overlapping the range of the Great Grey. Where these two species of grey kangaroo overlap in distribution, they do not interbreed. However where the large kangaroos overlap in this manner they tend to be separated by their distinct habitat preferences.

The Great Grey occupies a considerable diversity of habitats, including temperate woodlands, cool mountain forests, sub-tropical forests, coastal heathlands, and dry inland scrublands.

The Western Grey, which in South Australia and far western Victoria is known as the Mallee Kangaroo, Black-faced, or Sooty Kangaroo, has a preference for mallee and dense dry scrub habitats, as one of its local names indicates.

Over great areas of the western plains of New South Wales and Queensland the range of the Great Grey overlaps that of the Red Kangaroo by more than 400 km. However, the Red and the Great Grey are quite effectively separated ecologically, the Red preferring the open plains, and the Great Grey the forest strips near rivers and in valleys of ranges.

The Great Grey is one of Australia's largest kangaroos, with a head-and-body length of more than one and a half metres, and the tail another metre.

The Western Grey Kangaroo is a darker grey-brown than the Great Grey, and has a dark muzzle, and dark markings around the base of the ears. The big, very dark furred, slow-moving Kangaroo Island Kangaroo was the first-named variety of this species, that is, *Macropus fuliginosus fuliginosus*, the Black-faced, Sooty or Mallee Kangaroo is the race *M. f. melanops*, and the Western Grey of south-western Australia is the race *M. f. ocydromus*.

Red Kangaroo

Megaleia rufa

OF ALL the great kangaroos, the big Red has been the centre of greatest controversy and argument between those who, on the one hand, believe that the Red is a threat to the pastoral industry, and those who wish to see the species given greater protection. For this reason, the biology of the Red has been the subject of the greatest research.

The Red Kangaroo is the most widely distributed of the kangaroos, the only big marsupial over vast tracts of the dry inland plains. With its longer limbs the Red is ideally suited to the wide open spaces of grasslands, mulga scrublands, sandplains, and sand-ridge country.

The area of distribution covers the entire interior of the continent, extending to the coast in the north-west and at the head of the Bight. The limits to its habitat are set by the coastal woodlands and forests, which it avoids.

Within this enormous region the Red Kangaroo is most plentiful on the better country, plains of higher rainfall and more abundant herbage, shelter and water. It is in these areas that it is most in conflict with the pastoral industry.

Although occurring in desert areas, such as the Great Sandy Desert and Sturt's Stony Desert, the Red is generally quite rare in the most arid regions, so these cannot be of great value as sanctuaries for this species if it is ever completely exterminated from the better pastoral country.

Research has shown that normally the Red Kangaroo does not select the same food plants as sheep and, weight for weight, does not eat more than a sheep.

Kangaroo numbers must be controlled in some areas, and in such populations can be commercially exploited, but this must be done within an industry that is well regulated and based on sound scientific data on the biology of the kangaroo. In some States the uncontrolled slaughter of kangaroos has resulted in their extermination from many areas, and the pet food industry which has depended upon the kangaroo has had to suffer.

The kangaroo has great potential for protein production in semi-arid regions and with less damage to the environment than caused by sheep. The kangaroos are more efficient in conversion of grass to meat, with a carcass that is 52 per cent meat compared with 27 per cent for a sheep.

Not only is the habitat of the Red Kangaroo generally of rather low average rainfall, but it is subject to long and severe droughts, from which even the best parts of its range of distribution are not excepted.

The Red Kangaroo has the ability to withstand considerable extremes of heat and water shortage, although in this respect is probably second to the Euro.

During drought the kangaroos also cease to breed, as young are otherwise the first casualties.

Agile Wallaby

Wallabia agilis

COMMON across the grassland plains of tropical northern Australia, the little Agile Wallaby is one of the few smaller members of the kangaroo family that occurs in such abundance. In some localities at times many will be seen on and beside roads at night, and in the Northern Territory and far north of Western Australia this species has been declared vermin because of its numbers.

Recognition of the Agile will cause no difficulty in the natural habitat, because in its natural haunts there are few other wallabies of the grassland plains. In north-eastern Queensland, it overlaps the range of the Whiptail Wallaby, while across much of the north it coincides with the Northern Nail-tailed Wallaby. Each is quite distinctive in appearance, and even when only a fleeting glimpse is obtained, the sandy yellowish-brown or rusty-tinted fur of the Agile, together with its upright body posture when hopping, serve to separate it from the grey-furred Whiptail and the Northern Nail-tail with its horizontal head-low posture when travelling.

Also within the range of the Agile is the Antelopine Kangaroo, which has long shaggy fur and more thickset build, and of course adults are very much larger. There are also the rock-wallabies in the north, but these are generally segregated in habitat preference.

The Agile inhabits the tall-grass country of tropical plains and woodlands, with a preference for the river floodplain country. By day it hides in dense thickets, riverside scrub, or pockets of tropical monsoon forest, emerging towards dusk to feed on the open grassy woodland country.

Because of its very wide distribution across northern Australia the Agile has acquired a number of regional names, of which two are 'River Wallaby' and 'Jungle Wallaby'. As would be expected of an animal of such wide distribution, the Agile shows some differences upon which a number of regional races have been described. There is a north Queensland race said to be a brighter yellow colour, a Northern Territory race of a rather dull sandy-brown, and a race from the far northern parts of Western Australia with reddish-sandy upper parts. There is a very marked difference in size between males and females of this species.

Because the habitat of the Agile is not threatened, and because the species is common and already occurs in various northern national parks and reserves, it is at present not necessary to contemplate any special reserves for it.

The Agile wallaby was seen first by Europeans on the Coburg Peninsula, north-east of Darwin. The speed and agility with which the wallabies escaped hunters and dogs by dodging through mangroves, thickets, over logs and grass tussocks so impressed the early explorer-naturalists that they bestowed the *agilis*, agile.

On the east coast the species extends well down the coast, being found on Stradbroke Island off southern Queensland.

Woilie

Bettongia penicillata

THE RAT-KANGAROO group or sub-family contains the smallest and some of the most primitive of all the kangaroos. Apart from their small size and rather rat-like appearance, these marsupials are separated from other groups by their short rounded ears and more primitive arrangement of teeth. In this group, which includes some that are still partly insectivorous in the manner of the earliest kinds of marsupials, are included the bettongs and the potoroos. The bettongs are known also as brush-tailed rat-kangaroos. The potoroos are rather bandicoot-like, with comparatively short hind legs, and a galloping run rather than the usual kangaroo hop.

The larger kangaroos, wallabies and pademelons have teeth very much changed from the original primitive insectivorous dentition. For their purely herbivorous diet their front teeth are chisel-like, designed for snipping off grass or foliage. Behind is a broad toothless gap, then the herbage-grinding molars, broad-crowned but with sharp grinding ridges. The rat-kangaroos, however, retain typically canine pointed front teeth, more suited to a generalised rather than herbivorous diet.

The primitive ancestral type of kangaroo is probably best represented, among living species, by the Musky Rat-kangaroo *(Hypsiprymnodon moschatus)*, a rabbit-sized inhabitant of the rainforests of north-eastern Queensland. So different that it is generally placed in a sub-family of its own, the Musky Rat-kangaroo has a scaly, almost hairless possum-like tail, and other distinctive features.

Also separating rat-kangaroos from larger kangaroos are certain aspects of behaviour. Some and probably all species build nests of sticks, grass and other materials, while one being the only species of burrowing kangaroo, digs a burrow into which to place its nest.

The Woilie, also known as the Brush-tailed Rat-kangaroo, once inhabited most of the continent south of the tropics, but is now extinct in eastern Australia. Only in the south-west of Western Australia are there remnant populations in State forests, and other reserves which contain its preferred habitat of thick scrub undergrowth in dry eucalypt woodland.

The Woilie constructs a grass nest in a depression scratched out at the base of an overhanging bush or tussock of grass. This species has the remarkable habit of transporting nest materials in a bundle held by the curled tip of its tail, giving a most remarkable appearance hopping through the bush. While sleeping in its nest the Woilie pulls grass over the entrance, so that from the exterior there is no obvious entry, and the whole structure is most inconspicuous, looking rather like a pile of wind-blown debris. When an occupied nest is approached too closely the Woilie generally bursts out with explosive suddenness and has vanished into the scrub before the startled observer can even think of giving chase.

Tammar

Macropus eugenii

ALTHOUGH formerly found across large areas of the south-west of Western Australia and southern parts of South Australia, the Tammar or Dama Wallaby is now common only on some islands of the coast of those States.

On the mainland, the preferred habitat of this wallaby has been lost largely to wheat farming. The Tammar is an inhabitant of rather dry eucalypt woodlands where it shelters in dense thickets of mallee or scrubby undergrowth. The mainland populations of the Tammar are now isolated in remnant pockets, surrounded by the open paddocks of farmlands, which isolate one colony from another almost as effectively as if each was on an island surrounded by ocean. Many of these remnant areas of native bush are national parks or fauna reserves. At one of these, the Tuttanning reserve in south-western Australia, long-term studies are made of the Tammar and its preferred or optimum habitat, and the problems of preserving this habitat, particularly when most areas remaining are small, and subject to many outside pressures, such as weed-encroachment chemical applications to surrounding farmlands. Small reserves are also very vulnerable to fire, for there is a greater possibility that after a bushfire there will be insufficient area, if any, remaining for food and shelter of the wildlife.

Fortunately it is on the various offshore islands that the Tammar occurs in greatest numbers and has its greatest chance of survival as a species. The Tammar is common on Kangaroo Island, where it has the large Flinders Chase Reserve, and also on other islands of South Australia. Around the Western Australian coast it is abundant on some islands of the Houtman Abrolhos and Recherche Archipelago.

The Tammar is of scientific interest in that there are about eleven isolated populations. When the distribution of any species is broken up in this way, so that there is no contact and interbreeding between the groups over time intervals of thousands of years, each group tends to evolve differences, at first very slight, in response to the unique characteristics and pressures of its particular habitat. So the differences we see today, between populations on isolated islands, mark the early stages of the evolution of a single species towards a number of separate species. In the case of bush reserves on the mainland, comparatively recently isolated by destruction of surrounding habitat, the process has only just begun. The variations which gradually appear may be in colour, size or other physical features, or in behaviour.

As yet not a great deal has been done in comparative studies between Tammar populations, but some obvious differences were noted. The Kangaroo Island race is larger, more sturdily built, while those inhabiting the Abrolhos Islands and mainland south-western Australia have been described as other races.

The Tammar, in habits, is similar to many of the small pademelon wallabies in that it seeks refuge in dense low scrub, where it has a maze of tunnel-like runways. The species also shares a habit of many other gregarious kangaroos, warning of danger by a loud thump on the ground with the hind feet.

The Tammar differs from some other wallabies in that it inhabits some arid areas. Some of the islands are in very low rainfall regions. Recent studies have shown that it can survive without fresh water by eating plants containing water, some of the coastal succulents.

Bennetts Wallaby

Macropus rufogriseus

ONE OF A GROUP of nine macropods which could be classified broadly as large wallabies, intermediate in size between the really big kangaroos on the one hand, and the many small wallabies and pademelons on the other, Bennetts Wallaby possesses a confusion of common names.

Perhaps because it is found in a region which coincides with the bulk of the human population of Australia, it has acquired many localised district names, many dating back to the first settlement of each locality.

This wallaby tends to keep to the forest underbrush where it browses on foliage rather than venture out into open grassy areas, so in some places it is known as the 'Brush Wallaby', elsewhere as 'Brush Kangaroo'. Of similar derivation is 'Scrub Wallaby' of other districts. In eastern Queensland it became known as the 'Roany' for its more ashy brown colour in those parts. The name 'Red-necked Wallaby' is in widespread use applied to the wallabies of this species over a large part of their mainland range where there is a distinct reddish tone to the shoulders and back of the neck.

The name 'Bennetts Wallaby' was given originally to the Tasmanian race of this species, when that animal was first described as a species separate from those of the mainland (that is, the species *bennetti*). This Tasmanian wallaby, now regarded as no more than a sub-species, is smaller, but of more solid build, and more uniformly grey-brown, without such distinctive red shoulders when compared with the mainland examples of the species. The species was originally described from specimens taken on King Island in Bass Strait, and also occurs on Flinders and other Bass Strait islands. These specimens appear identical to those inhabiting Tasmania. The photograph shows the King Island form.

Because there are obvious colour differences between Bennetts Wallaby of these southern regions, and the east-coastal Red-necked Wallaby, these two names are in parallel use for the respective sub-species.

Bennetts Wallaby inhabits coastal scrubs and brush country from eastern Queensland through eastern New South Wales and southern Victoria to south-eastern South Australia (including the Red-necked) thence to Bass Strait islands and Tasmania. It is the only large brush wallaby in Tasmania, where it is often known simply as 'Kangaroo'.

Because this wallaby inhabits the heavier forests, brushes and scrubs, generally in rough country much of which is more or less in a natural state within forest reserves and national parks, and because it has such a wide distribution, the species seems secure well into the future. In Tasmania it is especially abundant, and inhabits some very mountainous country at altitudes of over 1,000 metres.

The large-wallaby group contains some nine species and the small-wallaby group about six species. A hind-foot length of 16.5 cm serves to divide the two groups. Smaller, and much lighter in build compared with the five species of great kangaroos, the large wallabies are much more graceful animals on the whole.

Bennetts Wallaby is now generally placed in the genus *Macropus* but was previously in the genus *Wallabia*. Studies of such factors as blood proteins and chromasomes have led to re-grouping of many species.

Southern Potoroo

Potorous tridactylus

SUPERFICIALLY RESEMBLING a bandicoot, the Southern Potoroo is a very small animal of the kangaroo family, having a head-and-body length of a mere 35 cm. Like a bandicoot it has very short rounded ears and rather short tapering prehensile tail. Compared with the larger kangaroos its claws are much more suited to digging and scratching about in soil and leaf litter.

As with many other species, not only of marsupials but of fauna and flora in general, there has been considerable name-changing. Some references list this animal as *Potorous tridactylus*, others as *P. apicalis*. The name to be used depends upon whether the Tasmanian potoroos and those of the islands of Bass Strait, southern Victoria and South Australia, all the southern regions, are to be regarded as a full species separate from the potoroo of east-coastal New South Wales and Queensland. If they are a separate species, then the potoroo shown in the photograph is *Potorous apicalis*, being a southern animal.

On the other hand, if the southern potoroos are merely a sub-species of the east-coastal species, then they, like the east-coastal species, are *Potorous tridactylus*, sub-species *apicalis*. To further complicate the matter, a potoroo which inhabited the south-west corner of Western Australia but which now is probably extinct, Gilberts Potoroo (*Potorous gilberti*), has been listed in some reference works as a distinct separate species in this manner. But more recent studies include this potoroo within the species *tridactylus*, that is, *Potorous tridactylus gilberti*. So current thinking is that there is a single species, *tridactylus*, with several sub-species. There is also another south-western Australian potoroo, the Broad-faced Potoroo, *Potorous platyops*, which has been left as a full separate species. This appears almost certainly to be extinct.

With common names, too, there is a wide choice. The southern animal is known as the 'Southern Potoroo', or simply as 'Potoroo'. The east-coastal animal as 'Long-nosed Rat-kangaroo' or as 'Potoroo', while in Western Australia there are 'Gilberts' and the 'Broad-faced'.

The potoroos remain most common in Tasmania, but are also quite common in some pockets of suitable habitat along coastal parts of eastern Queensland and New South Wales. This small member of the kangaroo family is particularly dependent upon the preservation of its natural habitat vegetation for survival.

Low damp localities where the herbage is very dense give shelter from the many potential predators. Destruction or thinning of the dense undergrowth leaves the little potoroos extremely vulnerable. A nest is built as a daytime hideout.

Potoroos are nocturnal and feed upon roots and tubers of plants. These are dug up with the long claws of the forefeet. Potoroos do not venture out of the dense vegetation to feed in open areas as do some small pademelons, bettongs and Woilie.

Like all other members of their family the potoroos carry their young in a pouch which has a forward-facing opening, unlike that of the superficially similar bandicoots, which have a rearwards-opening pouch.

Red Wallaroo

Macropus robustus cervinus

THE RED WALLAROO of north-western Australia, like its almost black-furred east-coastal counterpart, inhabits the rough ranges, breakaway and gorge country.

Equally well known by the name 'Euro', the Red Wallaroo, like others of the wallaroo group, is readily distinguished from other big kangaroos (the Red and the various species of grey) by the heavy thick-set build and long shaggy fur. When travelling, they tend to hold their bodies in a very stiffly upright posture. In limb proportions too there are differences, their hind limbs being shorter, in proportion to their bodies, than those of the other big kangaroos.

There are of course many other features distinctive of the wallaroos, some evident only upon close examination, such as the teeth, and anatomical differences of skulls.

A ready aid to identification of the various big kangaroos is the snout tip. The nose tip of the wallaroos has a large area of bare skin, the greys very little, and the Red intermediate.

The wallaroos, rather like the smaller rock wallabies, inhabit rocky and hilly places, and could well be called rock kangaroos or hill kangaroos.

An old male wallaroo is a big and heavy built animal, appearing more muscular and powerful than the Red or greys. If cornered in its rugged retreat it can be dangerous to approach closely. It is well able to use claws and teeth. Old males of the wallaroo tribe seem typically ill-tempered, which makes them dangerous as pets or elsewhere in captivity.

The kangaroos of the wallaroo-euro group are found over a very large part of the Australian continent, principally inland, but also in some coastal districts. Their habitats include forested ranges of the south-eastern coast, rugged sandstone plateaus of tropical Arnhem Land, and deserts of the centre. Their distribution appears not to be determined by climate, but by availability of suitably rough habitat. In the arid regions the wallaroos populate the lateritic breakaways and stony hills. In coastal districts, they go to the rocky crests and ridges of forested ranges. From these secure hideouts they often will descend to surrounding grassy flats and open forest to feed in the evenings and early morning.

The euro is extremely well adapted to survive the frequent droughts of the arid parts of the environment. By avoiding the heat of the day, seeking out the cool caves and crevices of rocky ranges, feeding during cooler hours of the day, seeking out the most succulent and water-holding of the plants, and by their ability to tolerate considerable water loss, they can inhabit the hottest and driest areas, although only the hardiest survive the really severe droughts. With the provision of water for domestic stock, the euros of the dry regions have become more numerous, but those of coastal regions in most places have diminished as a result of destruction of their natural habitat.

Roan Wallaroo

Macropus robustus erubescens

ONE OF many sub-species of the stocky big kangaroo known as the Wallaroo or Euro, the Roan Wallaroo is a distinctly coloured form inhabiting central Australia and South Australia. The hairs of its long shaggy reddish-toned fur are, on its upper surfaces, black-tipped, giving the roan colour effect. Its colour tends to be more or less intermediate between the dark charcoal of the east-coastal Wallaroo and the very red Euros of north-western Australia.

Although in the past described as a separate species of great kangaroo (the species *Macropus erubescens*), present classification has relegated the Roan to mere sub-species level (*Macropus robustus erubescens*). It is distinguished from other euros and wallaroos, with which it is one and the same species only by extending to the three-name, sub-species, level.

To add to the confusion, some older books had placed all species of wallaroos and euros in a separate genus, *Osphranter*, and listed this particular animal as *Osphranter erubescens*.

It is now thought that some forms of wallaroo presently given sub-species distinction should not even be given that level of difference. They are but localised variations not sufficiently different to deserve a sub-species name. The scientific reasoning here is that the colour variation, from almost black at the east coast, through increasingly reddish tones to deep red-brown at the north-west coast, is a continuous gradation. There is probably nowhere an abrupt step in the colour scale, so that it becomes difficult to recognise definite sub-species boundaries based on fur colour.

In such a graded range of a feature, known as a 'cline', there is no point at which it is possible to say that the realm of one sub-species ends and that of another begins.

But where there are geographic breaks in distribution, due to a physical barrier, distinct sub-species and eventually full species evolve. On Barrow Island, off the north-west coast, the Barrow Island Euro (*Macropus robustus isabellinus*) is evidently reasonably distinctive. In this case, the male is buffy red.

Like most other wallaroos and euros, the Roan has a preference for ranges and rocky country. Only one member of the wallaroo group is an inhabitant of grassland plains, and that animal is not known as a wallaroo or euro, though the resemblance can be seen. This is the Antelopine Kangaroo, which on tropical northern Australian plains and savannah woodland country takes the place of the Red Kangaroo and greys of more southerly plains.

In those northern parts, the Antelopine (*Macropus antilopinus*) overlaps the distribution of the Northern Wallaroo (*Macropus robustus alligatoris*) but each keeps to its own habitat, the wallaroo to the rough ranges, the Antelopine to the grassy flats.

The reddish-tan coloured Antelopine does not share with wallaroos the typical upright posture of the body while bounding along, but adopts a more horizontal body line, and could easily be mistaken for a Red Kangaroo when bounding across the plains. It is rather long-limbed and probably as big as many large Reds, but when seen close at hand it has a rather wallaroo-like face, and most noticeably the long shaggy wallaroo fur unlike the very short fur of the Red.

Yellow-footed Rock-wallaby

Petrogale xanthopus

AUSTRALIA has seven species of rock-wallaby, grouped in two genera. Their most obvious characteristic in common is their specialisation for mountainous rocky country, where they shelter among the boulders and caves by day, and feed on nearby flats towards dusk and very early in the mornings.

But it is their incredible agility on the cliffs and boulders of their rugged hideouts that is the forte of these kangaroos. With nonchalant ease, grace and precision a rock-wallaby will leap from ledge to ledge across a cliff-face, from boulder to boulder across deep crevices, hurl itself down a precipitous slope and within seconds be lost from sight.

For this way of life the rock-wallabies have evolved adaptations which make possible the breathtaking acrobatics which put them far ahead of any pursuer in their natural rugged haunts.

When airborne in mid-leap, maintaining balance on landing, and making abrupt changes in direction, the rudder and counterbalance effect of the tail is vital. For greatest effect, therefore, the tail is long, comparatively heavy, cylindrical more than tapering to put more counterbalancing weight near the end, and often bushy towards the tip to give a greater rudder-like effect through increased wind resistance during airborne stages of the action, which is, with a rock-wallaby at speed, probably most of the time.

The hind feet of the rock-wallaby have thickened pads, soft, yet with a granulated rough surface, for a non-slip grip on smooth rock.

Although ever alert and often able to vanish without being seen at all, rock-wallabies leave evidence of their presence in the polished rock surfaces of ledges and cave floors where their traffic has been concentrated over many thousands of years.

Rock-wallabies are found over a very wide range of environments from rocky coastal islands to central Australian desert ranges. In the hot arid regions their habit of resting throughout the day in cool shady and sometimes moist caves is a behaviour adaptation which conserves body moisture and shelters them from the heat of the day. This possibly has permitted them to inhabit regions that otherwise would be climatically too severe for them.

These wallabies are gregarious, in suitable places forming large communities with social behavioural patterns which undoubtedly will provide interesting sociological studies in the future.

Of the seven species, one is especially widespread. This is the Brush-tailed Rock-wallaby, which has a far-flung distribution across the mainland, and to some offshore islands. The island groups are isolated, as are mainland colonies on ranges separated by flat country, though to a lesser degree. Some are recognised as distinctive localised races, such as the Pearson Island Rock-wallaby, the Recherche Islands Rock-wallaby, the Western Rock-wallaby, all being but slightly differing forms of the Brush-tailed Rock-wallaby.

Nowhere near as widely distributed, the Yellow-footed Rock-wallaby is found in the Gawler Range and Flinders Range of South Australia and arid interior ranges such as the Grey Range and the Main Barrier Range of western New South Wales, and south-western Queensland. Its ringed tail, yellow hind feet and forearms, and large furry yellow-and-white ears are distinctive.

Six of the rock-wallaby species are grouped in a single genus, while a single species the Little Rock-wallaby, forms the other genus.

Red-bellied Pademelon

Thylogale billardierii

THIS SMALL kangaroo is extremely common in many parts of Tasmania and occurs also on some Bass Strait islands, but is now very rare in places where it was once abundant in southern Victoria and South Australia.

The Red-bellied Pademelon is a gregarious species, living in large communities in gullies, around swamps and other low damp places where the vegetation is low, dense and tangled. Here it forms tunnel-like runways, a maze of interconnecting passageways where the many little pademelons of the colony can lose most pursuers.

In Tasmania this pademelon is to be seen in many of the national parks, most often in some of the camping grounds of mountain national parks such as Cradle Mountain and Lake St Clair where they may become quite tame.

Where not hunted or otherwise frightened, the little Red-bellied Pademelons emerge in early morning and towards dusk to feed in grassy clearings, but are always quick to dart back into the shelter of the dense vegetation at the first sign of danger.

While a distinguishing feature of this species is the fur coloration, yellowish or dusky orange below, dull brown above, there is considerable variation. In colder and wetter regions including the Tasmanian highlands, the coat is more shaggy and sombre coloured.

Outside the national parks and similar reserves the abundance of the Tasmanian pademelons poses a problem. It is claimed that on farmlands their numbers cause a reduction in farm production, while in some forestry areas, where seedling softwood trees are being established, the pademelons can cause quite heavy losses of young trees.

From time to time, as a control measure, large numbers of the pademelon in Tasmania are shot, while a great many others are taken for their fur. Some two-and-a-half million skins were taken in Tasmania in the years 1923 to 1960.

Balancing the claimed economic damage, the presence in numbers in some places of this delightful small pademelon is a unique feature of Tasmania which adds much to the character of its mountain parks.

There is little difference between those of the small kangaroos known as 'pademelons' and others slightly larger given the name 'wallaby'. However pademelons are generally noticeably smaller, with a hind-foot length of less than 15 cm.

As with so many other mammals, there has been considerable reclassifying of the pademelon wallabies in recent times as evidence of a scientific nature reveals similarities or differences not previously suspected. There are now three species of pademelon of the genus *Thylogale*, where in the past there were as many as eight species. Some former *Thylogale* wallabies have been shifted to other genera, for example, the Tammar, formerly *Thylogale eugenii*, is now *Macropus eugenii*. Other species have vanished, now being considered merely slightly differing races not worthy of full separate species status. The Southern Red-legged Pademelon (*Thylogale wilcoxi*) and the Cape York Pademelon (*T. coxenii*) of older reference books have been combined now with the Northern Red-legged Pademelon (*T. stigmatica*) and all are known by the one common name, Red-legged Pademelon, and the species name *stigmatica*. So there is a trend towards simplification, a reduction in the past confusing number of species, but confusion can arise if modern references are compared with older works.

Spectacled Hare-wallaby

Lagorchestes conspicillatus

THE NAME 'hare-wallaby' was given when, in the early days of the exploration and settlement of Australia, these rabbit-sized wallabies were abundant.

If their nesting places under tussocks of grass or low bushes were approached they would dash out with hare-like abruptness and speed. However, the species which inhabited eastern Australia is probably extinct, the last having been seen in 1890; this was the Brown Hare-wallaby, *Lagorchestes leporoides*.

Three species still survive because they occur in remote areas, on islands which give them protection from the introduced cat and fox. The Banded Hare-wallaby, *Lagostrophus fasciatus*, is common on Bernier and Dorre Islands in Shark Bay, Western Australia, but no specimen has been taken on the mainland since 1906. The Western Hare-wallaby, *Lagorchestes hirsutus*, occurs on these same islands, and inhabits spinifex grasslands in the far northern interior of Western Australia where it may still survive in the remote country traversed by the old Canning stock route.

Like these other island refugees, the Spectacled Hare-wallaby has its stronghold on an island, Barrow Island, off the north-west coast of Western Australia. It also occurs in the remote northern parts of the interior of Western Australia, Queensland and the Northern Territory. It is probably the only member of the once abundant hare-wallabies still reasonably secure on mainland Australia. Although all species have orange rings of fur around the eyes, the feature is far more noticeable on this species, hence the name 'spectacled'.

The original specimen of this animal was collected on Barrow Island by the naturalist Byrnoe of the survey vessel HMS *Beagle*, and sent to the British Museum. It was named by Gould in 1842, the species name *conspicillatus* referring to the reddish patches encircling the eyes, which give the spectacled appearance.

A mainland sub-species was discovered by Leichhardt on his exploratory expedition from Moreton Bay, Queensland to Port Essington, Northern Territory, probably captured between Normanton and Darwin. This sub-species was said to be larger, more colourful; it was named, also by Gould, in honour of Leichhardt, *L. conspicillatus leichhardti*. A north Queensland sub-species, *L. c. pallidor*, is, as its name indicates, of lighter colour.

Although the distribution of the Spectacled Hare-wallaby has been greatly reduced, there remains a population around Charters Towers in Queensland, and this is probably the last population of any kind of hare-wallaby surviving in eastern Australia.

There may still exist, although this is most unlikely, another species of *Lagorchestes*. A single skull was collected near Lake Mackay, on the Western Australian-Northern Territory border, in 1932 or 1933. Features of this skull are sufficiently distinctive to establish it as belonging to a separate species which was given the names *L. asomatus*, the Desert Hare-wallaby. No other evidence of this animal has yet been reported.

Whiptail Wallaby

Macropus parryi

UNLIKE MANY of the smaller members of the macropod family, the Whiptail Wallaby remains, for the present, secure as a species. In part this must be as a consequence of the habitat preferred by this wallaby: hilly country with open woodland, grassy beneath rather than scrubby.

Although the range of this species has been significantly reduced in those areas where this type of environment has been destroyed on a large scale, there has remained a great deal of this kind of country especially along the Great Divide, coinciding with the Whiptail's range of distribution. The Whiptail inhabits eastern Queensland and north-eastern New South Wales. Not surprisingly, it is in the latter region, where land exploitation has been greatest, as in south-eastern Queensland also, that this species has been most reduced in numbers. But although a considerable area of habitat remains, more national parks and other secure reserves will be needed to assure the Whiptail's long-term survival.

The country inhabited by the Whiptail, open-forested slopes and summits, is commonly utilised by man without major change to the environment, principally for the purpose of livestock grazing. While the valley flats and lower slopes are cleared, the higher, steeper parts are left uncleared very often, and livestock graze among the trees. There is probably a reduction in undergrowth other than tall grass, but pasture improvement and provision of watering points often mean an improved habitat for the wallabies, unless there is an attempt to eliminate the wallabies by shooting or other means.

Other extensive areas are forestry reserves and, again, while there are changes brought about by logging, the habitat as a whole usually remains suitable for these wallabies. Exceptions are where there is extensive clear-felling, and the planting of pines, which are shunned by most native animals.

The Whiptail, now carrying the scientific name *Macropus parryi*, but also in the past described as *Wallabia parryi* and *Wallabia elegans*, is one of the larger of the wallabies.

It is also known in common-name terms as the Pretty-face Wallaby, Grey-faced Wallaby, Blue Flier and Grey Flier. In scientific nomenclature there have also been a number of names. Older references separate the present single species into two, *Wallabia cangaru*, and *Wallabia elegans*. The former, given the common name of Captain Cook's Wallaby, covers the northern Whiptail population, while those southwards of Rockhampton were, as now, the species *elegans*, the Whiptail. The southern animals were said to be sufficiently different from the northern by virtue of their larger size and paler fur, with more strongly contrasting facial markings.

The most distinguishing feature of all members of the present single species is the very long slender tail, which is as long as the combined length of head and body. There is also a very conspicuous white face stripe, and a noticeable white hip stripe, which extends under the tail from one side to the other.

Whiptails are said to be faster moving than most other wallabies, due no doubt to their more open habitat where speed rather than concealment is the key to survival. Where not excessively disturbed they may be seen feeding late into the morning and again during the afternoon. They seem more gregarious and more diurnal than many of the wallabies of scrubbier habitats.

Common Ringtail Possum

Pseudocheirus peregrinus languinosus

THE Common or South-eastern Ringtail is the well-known ringtail possum of a region extending from Adelaide, around the south-eastern corner of the continent, and northwards to the south-east of Queensland. Its range further west is restricted by the lack of suitable forested habitat west of South Australia's Mt Lofty Ranges. In more northerly parts of Queensland it is replaced by the Queensland Ringtail and the Rufous or Bunya Ringtail. This is probably the best known of Australia's ringtail possums, for its range covers the major cities and it may be found in gardens of some suburbs.

Although a number of formerly separate ringtail species are now regarded as one, it is desirable to retain their various common names, for these animals are noticeably different especially in fur colour.

This form of Common Ringtail of the south-east corner was formerly the species *Pseudocheirus languinosus*, as named by Gould in 1858. However the Queensland Ringtail, which seems to have been the form collected by Captain Cook, at the site now occupied by Cooktown, in 1770, was in 1785 named *Pseudocheirus peregrinus*. The rule applies that the earliest name is to be used, so now that many previously separate species have been lumped together, all must go by the species name *peregrinus*. But if being discussed at the sub-species level, and if scientific names must be used, then the former species names become the third part of the sub-species names, i.e. the Common Ringtail, *Pseudocheirus peregrinus languinosus*, the Queensland Ringtail, *P. p. peregrinus*, and the Western Ringtail is *P. p. occidentalis*.

When Cook discovered that first ringtail in 1770, marsupial opossums were already known from North America. He recognised the similarity, and entered in his journal a description of a 'possum', omitting the first 'O'. This shortened name has been retained, and ever since has served to separate the Australian possums from American opossums, and such differentiation is useful because the two kinds are but distantly related.

One of the obvious anatomical features of the possums is the long prehensile tail. The origin of the name 'ringtail' is obvious when it is seen how the tail is rolled into a ring, or more correctly a neat tight spiral when not being used. It is generally curled up in this way as the possum enters its nest or hollow to sleep for the day.

When climbing in the treetops the ringtails have a graceful, deliberate action, making more use of the powerful hand-like hind feet and the encircling grip of the tail than of the small front feet. Crossing from one tree to another can be accomplished even when the trees are not quite touching, the possum hanging out from the swaying tip of a supple branch to catch with forepaws the foliage of the second tree. The tail grips as a 'safety rope' until the possum has swung itself across. Ringtail possums, unlike smaller species which are partly insect and nectar eaters, are vegetarians, and feed on the foliage of eucalypts, peppermints and other native trees and shrubs.

Apart from these ringtail possums of the genus *Pseudocheirus* there are two other ringtail genera. The Rock-haunting Ringtail *Petropseudes dahli* of the Northern Territory and the Kimberley has a short and apparently non-prehensile tail, while in the rainforests of the Atherton Tableland in Queensland, there is a rather strange looking Brushy-tailed Ringtail, *Hemibelideus lemuroides*, or Lemur-like Ringtail.

Honey Possum

Tarsipes spencerae

THE SOUTH-WESTERN corner of Western Australia is famed for its wealth and variety of flowering plants. Many of Australia's most spectacular and most unusual wildflowers are confined to this region. Whole families and genera are exclusively south-western.

Not all south-western habitats share equally in this wealth of wildflowers. The coastal sandplains are among the best endowed, and the sandplains around the south-western coastal strip are certainly not inferior to any other region.

The south-coastal sand-heath lands include swamps, banksia woodlands, eucalypt woodlands, granite hill slopes, and low windswept seaside sand heath country.

Large-flowered plants of this region include the Scarlet Banksia (*B. coccinea*), Baxter's Banksia (*B. baxteri*), Red Swamp Banksia (*B. occidentalis*), Albany Bottlebrush (*Callistemon speciosus*), Red-flowering Gum (*E. ficifolia*), Swamp Bottlebrush (*Beaufortia sparsa*), Showy Banksia (*B. speciosa*), Southern Plains Banksia (*B. media*), Warted Yate (*Eucalyptus megaconutu*), Prickly Honeysuckle (*Lambertia echinata*), Chitlick (*L. inermis*) and Barrens Regelia (*R. velutina*). A comprehensive list, even if including only large, showy species, would be much too long to include here.

These wildflowers have not evolved in isolation. The flowers serve a purpose, that generally being to achieve effective cross-pollination. For this, many rely upon nectar-eating birds, as well as, or in place of, the usual insect pollinators. But this wealth of large flowers, with a copious supply of nectar to attract birds, and with various species flowering in different months right through the year, has permitted the evolution in this region of a unique little marsupial, one which is entirely dependent upon these wildflowers for its existence. This is the Honey Possum, a mouse-sized maruspial which occurs in no other part of Australia or elsewhere in the world. So dependent is it upon nectar that its specialisation has gradually caused it to develop a long, slender tubular beak-like snout, with a tiny round mouth at the tip, and a brush-tipped tongue like those of the honey-eating birds. The teeth are greatly reduced and degenerate, so that this little animal is now tied to the liquid diet of nectar. But without doubt it would, in with the nectar, obtain many tiny insects, of almost microscopic size, which often occur in thousands in some banksia and other flowers, and probably pollen also.

The unique little Honey Possum is confined to the south-west corner from the Murchison River to Esperance, usually on sandplain country. Although well equipped for climbing, it seems to spend much time on the ground where it is fast, quite unlike the slower moving pigmy possums. It seems also to be at least partially diurnal in habits, there being a number of reports of sightings of the Honey Possums feeding or active in full daylight.

The Honey Possum is said to build a nest of grass and fur, woven into dense foliage, or to hide in an abandoned nest of a small bird. The young are probably kept in the nest for a considerable time after leaving the pouch.

This unique little marsupial appears to have no close relatives among the marsupials that inhabit this continent today. Its origins are to a large extent concealed by its special adaptations. It is therefore placed in a family of its own, the Tarsipedidae.

Herbert River Ringtail

Pseudocheirus herbertensis

ALTHOUGH bearing the common name of Herbert River Ringtail, this possum in fact has a somewhat wider distribution than the region around that river. This species was discovered by Dr Carl Lumholtz in the rainforests of the highest parts of the north Queensland ranges at the headwaters of the Herbert River, inland from Innisfail. It was later found to inhabit high altitude parts of the mountain chain from Townsville northwards to Cooktown.

The Herbert River Ringtail is one of the most boldly coloured of Australian ringtail possums. The fur is for the most part a deep chestnut brown, appearing black in the dimly lit rainforests of its mountain habitat. There are usually, but not invariably, conspicuous patches of white. This white fur is predominantly on the undersurface, where it shows along each side of the body when the animal moves along the tree limbs. The white commonly extends onto the arms—like white sleeves. These white areas show considerable variation. The chest and belly may be entirely white, other animals may have these underparts greyish, or broken into several white patches. In early days it was thought there were two species, a brown, and a white-armed.

In overall effect the colour pattern is boldly attractive. At close range the effect is greatly heightened by the red eyes, which bulge outwards in hemispherical shape the way the eyes of most nocturnal marsupials do, looking like glowing coals dropped onto dark brown carpet.

There are several slightly differing races of the Herbert River Ringtail. One form from Mt Spurgeon, about 100 kilometres north-west of Cairns, is a beautiful pale caramel tone. So restricted and inaccessible is its habitat (rainforest above 1,000 metres altitude) that this species was not discovered until 1937. This form, *Pseudocheirus herbertensis cinereus*, is uniformly coloured a subdued caramel-brown, without white markings but with a dark stripe down its head. There is also record of a lowland swamp form near Cairns, but not so obviously different in appearance from the usual Herbert River Ringtail.

Like the Green Ringtail, the Herbert River species is an example of an animal whose distribution is directly related to altitude, being generally restricted to the mountain tops over about 800 metres altitude. Australia has few ranges of height sufficient to greatly alter habitats and control speciation, but New Guinea has a great many examples of species isolated on mountain tops and individual valleys.

Herbert River Ringtails even today are not well known in details of biology; it is known that they feed upon leaves and fruits of such highland rainforest species as the Candlenut tree and the Sarsaparilla tree. Two young form the usual litter, and these are first seen out of the pouch around mid-October. When too large for the pouch they ride clinging to their mother's back, and begin an independent life when about half-grown. The young at that age have not yet developed the strongly contrasting deep-brown-and-white, but are a pale gingery colour on those parts which will eventually be almost black, and a buff tint in areas that later will become white.

Green Ringtail

Pseudocheirus archeri

THE ATHERTON Tableland, in north-eastern Queensland, is one of the richest areas within Australia for mammals, especially the mammals of the rainforest environment.

Situated in a region that receives an exceptionally high average annual rainfall, the Atherton Tableland of Queensland is surrounded by mountain ranges. To the east is the Bellenden Ker Range, of which the summit, Mt Bartle-Frere, is the highest peak in Queensland. Northwards are extremely rugged ranges around Mt Spurgeon, near Mossman, and southwards the mountain country of the headwaters of the Herbert River.

These coastal ranges, with steep slopes, precipitous ridges, deep gorges and high waterfalls, are clad largely in dense rainforest.

The unique nature of the fauna of the Atherton Tableland and surrounding mountains is due in part to the high altitude, which changes significantly the character of the vegetation. Rainforests cover, or did cover until cleared, much of the lowland plains, and extend upwards to the thousand-metre level of the tableland. The flora and fauna of the lowland rainforests is quite different to that of the highlands. A number of marsupials occurring on the Atherton Tableland and surrounding mountain ranges are found nowhere else in Australia. Other species occur further north, in the mountain rainforests of the eastern edge of Cape York, the mountainous country around the headwaters of the Daintree River, and extend only as far south as the Atherton or Herbert River districts.

The Green Ringtail is one of the species found only in the high-altitude rainforests of the Atherton Tableland, Herbert River and surrounding mountain ranges. It is a pale golden colour with a greenish tint, and has two stripes down its back. The green tone is not strong, and depends to some extent on the nature of the ambient light. In the natural green-toned forest surroundings and green-filtered light, it would be more green than in sunlight. From this strong light the shy nocturnal animal normally conceals itself.

The greenish tone is most unusual among furred animals, and is a combination of pigment and structural effects that blend together black, yellow and white. This possum is closely related to a green mountain ringtail of New Guinea. It is no doubt one of those creatures which invaded north-eastern Australia from New Guinea, and since then has remained confined to a region most like its original environment.

The Green Ringtail, also known as the Striped Ringtail or Toolah, is one of the most attractive of possums, with very big bulging hemispherical bright brown eyes, and white marking about the eyes and ears.

Mountain Brushtail Possum

Trichosurus caninus

THE NUMBER of species of brushtail possums inhabiting Australia has been amended on almost as many occasions as there have been zoologists studying this group of marsupials. Old reference books describe a great number of species, one for almost every locality, each slightly different in colour, size or other character.

Today, as a result of more precise scientific methods, the number of true species has been reduced to three, these being the Common Brushtail Possum, *Trichosurus vulpeca*, the Mountain Brushtail or Short-eared Possum, *T. caninus*, and the Northern Brushtail, *T. arnhemensis*.

Of these three species, the Mountain Brushtail and the Northern Brushtail inhabit relatively restricted regions or habitats, but the Common Brushtail ranges across the entire continent except for the Kimberley, the 'top end' of the Northern Territory, and Barrow Island. It therefore is found in a tremendous diversity of habitats, from deserts to alpine forests. There are probably about seven sub-species of the Common Brushtail, these including the Coppery Brushtail of Queensland, the Tasmanian Brushtail, and the South-western Brushtail.

The Mountain Brushtail is, however, not merely a sub-species of different colour, but is a full separate species, distinguished by its much shorter and more rounded ears, and tail that tapers more conspicuously because the fur is bushy towards the base and short towards the tip.

This species occurs in eastern Australia, from southern Queensland to Victoria, in heavily forested coastal mountain ranges, most often in rainforests.

The Mountain Brushtail must be one of the most attractive of the large possums, with big bulging light brown eyes more like those of a ringtail possum than of a brushtail, and a face that carries an alert, almost a permanently surprised expression. The fur is luxuriant with a silvery sheen.

Unlike the Common Brushtail, which is a very common animal in and about the suburbs of the cities, living in house roofs and generally benefiting from the changed environment, the Mountain Brushtail is much more the wild shy animal of the most dense forests, and one which would almost certainly vanish whenever its habitat is damaged or adversely disturbed.

The Mountain Brushtail, like all other possums, has very hand-like hind feet, adapted for climbing. Possums are vegetarian, feeding on foliage and fruits, and have the diprotodont type of teeth, with a single pair of broad lower incisors used for pulling away the vegetation.

Although a single young is usual among brushtailed possums, twins and even triplets have been recorded.

Brushtail Possums are responsible for some damage to regenerating forests, and in Victoria both the Common Brushtail and the Mountain Brushtail cause damage in pine plantations.

Male brushtails establish considerably larger territories than the females, so that the territory of each male overlaps the territories of several females.

Greater Glider

Schoiobates volans

LARGEST of the glider possums, the Greater Glider has a body about the size of that of an ordinary cat. The fur is shaggy dark grey or black. The tail is long and fluffy-furred. Some individuals are pure creamy white or very pale grey. Some combine the very dark and very light colours.

The Greater Glider is an inhabitant of eucalypt forests and woodlands of eastern Australia from the Dandenong Ranges to the Rockhampton area of eastern Queensland. It occurs also as a separate population in north-east Queensland. This is the sole species of its genus and of all the gliders is furthest adapted towards the herbivorous way of life. It is more closely related to the ringtail possums than to the other gliders and shares with them the leaf-eating preference.

The Australian gliders have evolved from three different groups of possums and are thus more closely related to those groups than to each other. The Feathertail has evolved from or closely in parallel with the Pigmy Possum and is placed with it and the Burramys Possum in the family Burramyidae. The Sugar, Squirrel and Fluffy or Yellow-bellied Gliders are as a group most closely related and similar to the rare, non-gliding Leadbeater's Possum. The Greater Glider as mentioned already is almost a fluffy flying ringtail.

That each of these groups seems to have independently evolved very similar gliding membranes (as have the overseas flying lemurs and squirrels) demonstrates the effectiveness of this means of travel and escape in a forest habitat. It is an interesting example of convergent evolution.

Although agile in the treetops and able to perform graceful glides of up to 150 metres, the Greater Glider is painfully slow and clumsy on the ground. It is an easy prey for its principal enemies the Powerful Owl and fox, should it misjudge a leap and land on the ground instead of on a treetrunk.

The Greater Glider chooses a large hollow high up in a tree and may line it with leaves and bark. Generally only one young is reared at a time. By the age of four months the young has outgrown the confines of the pouch. It clings to its mother's back. Presumably she becomes accustomed to this added burden which, although negligible at first, grows greater only gradually. The limitations of a steeper glide path and shorter maximum flight distances would be imposed gradually and be accepted as normal.

Australia has one other species of glider, apart from those shown, the Fluffy or Yellow-bellied Glider, *Petaurus australis*, also an inhabitant of forested coastal eastern Australia. This species, grey-brown on the back, is on its underside creamy white to rich orange, and has a total length of about 75 cm. As its generic name indicates, it is related to the Squirrel and Sugar Gliders.

Squirrel Glider

Petaurus norfolcensis

ON THE MANY continents around the world there has been a tendency for animals of differing origins to solve problems in distinctly similar ways. Similarities of habitat, a common problem such as the need to evade predators, have led very dissimilar animals, perhaps over long periods of evolution, to develop solutions remark-remarkably alike.

This process whereby animals, initially unlike, become at least outwardly alike through environment pressures is known as convergent evolution. There are many examples. The Australian Platypus, a primitive monotreme, has come to resemble the beaver, or vice versa. They occupy similar positions within their respective ecosystems as mammals dependent upon lakes or streams. The Australian spiny anteater or echidna, a monotreme, has a spiny armour which has almost exactly been duplicated by the hedgehog, a European placental mammal.

Similarly, in forests of tall trees and open air spaces, flight from tree to tree is far safer than across the ground for an arboreal animal. Although true flight is the province of birds, bats and insects, a compromise has been developed. Both mammals and reptiles have evolved gliding species. In tropical rainforests, although not in Australia, there are tree-dwelling lizards able to descend on a controlled downward glide. Several groups of arboreal mammals have developed gliding as an efficient and safe means of travel from tree to tree. A very effective means of escaping a predator encountered in the treetops must be to leap from the tree, and seconds later land on the trunk of another tree 50 or 100 metres away.

This ability has been evolved by both the North American flying squirrels, and by the flying lemurs of Malaysia and the Philippines. But in Australia, gliding mammals have independently evolved in the more primitive marsupials—the glider possums.

The similarities of the problem, the forest habitats, and the basic likenesses between mammals, have led to remarkably close appearances of these gliding mammals. Their flight surfaces are broad side extensions of the body skin attached between fore and hind legs, giving a rectangular parachute-like shape when all four limbs are outstretched. A long fluffy tail gives aerial control in a rudder-like manner.

The Squirrel Glider appears almost identical to the Sugar Glider, but is larger, and has a much more restricted range of distribution. While the former is found around northern and eastern forested coastal areas, the Squirrel Glider occurs only from eastern Victoria northwards to eastern Queensland.

Visually the main differences between the Squirrel Glider and the Sugar Glider are that the former is some 50 cm in total length against less than 40 cm for the Sugar Glider, and that the Squirrel Glider has belly fur white or creamy white instead of pale grey. To confuse matters, the Sugar Glider, in its wide range, varies in size, a female from the Drysdale River of the Kimberleys of Western Australia being far larger than a particular female of this species caught at Mareeba, north-east Queensland.

Sugar Glider

Petaurus breviceps

OF THE gliding possums the Sugar Glider is the most widely distributed and one of the most common. This species has been studied quite intensively.

The Sugar Glider in common with other Australian gliders is nocturnal in habits. After emerging from its nest after dusk it will forage among the treetops, often in company with others of its family group. A wide variety of feed is acceptable. The nectar of flowering trees is eagerly sought, and soft fruits are eaten. Insects are irresistible. The glider will leap at any insect which betrays its position by movement, seize it in its forepaws, and without delay bite its head repeatedly. Grasshoppers, moths and large spiders are all quickly killed and eaten.

Sugar Gliders will tackle quite large prey. One of five caught during a biological survey of the Drysdale River National Park in the Kimberley region of Western Australia was eating a Peaceful Dove when picked up in the beam of the spotlight.

These gliders, like most possums, have a variety of calls. A sharp bark-like yap which may be a curiosity or warning call, audible from some considerable distance, and a harsh rasping sound when angry. This angry noise, when issuing from, and greatly magnified by, a hollow of a tree, is most alarming. It is remarkably like the sound made by young kingfishers in their hollow tree nests. In both cases it must have considerable deterrent effect upon any predator, especially when the animal making the sound is hidden in the darkness of the hollow and is of unknown size and danger. The angry sound of the Sugar Glider is usually given also when biting during fighting or defending itself.

The Sugar Glider is a gregarious species, a group often consisting of parents and one or two generations of young occupying the one hollow. Within the group a social order of dominance is established. There are rarely serious fights but some occasional minor bickering. Scents play an important part in the social organisation. While each animal has its own individual smell, the group as a whole apparently becomes identified by the scent of a few 'high-ranking' males. The territory around the home tree covered by the group also becomes identified as tree trunks and other frequented vantage points become marked with this scent.

Although accepting other members of their own group, Sugar Gliders will fight with intruders from other groups, and have been known to kill a strange glider released into their territory.

The Sugar Glider travels through its forest or open woodland territory by leaping from a high part of a tree. It drops steeply at first, but with increasing speed and lift under its outstretched canopy the dive flattens to 30 or 40 degrees from the horizontal. In the final seconds of the flight the glider swoops upwards, its body becoming almost vertical for an instant to break its speed before it either hits a lower tree trunk almost simultaneously with all four outstretched limbs, or drops into the grass, if no tree is within gliding distance. A distance quoted as normal maximum flight is 46 metres, but much depends on surroundings and circumstances. I have seen one of these gliders. launching from quite a low tree but in hilly country, disappearing beyond the beam of the spotlight far downhill, probably higher above ground than it started. It could have gone as far as the range dropped steeply.

Feathertail Glider

Acrobates pygmaeus

THE SMALLEST by far of the gliders, the Feathertail or Pigmy Glider is about the size of a small common house mouse. It is very similar in size and appearance to the Pigmy Possum, of which it could be described as a flying version. It differs from the Pigmy principally in having the typical glider skin flap between front and back legs, and a feathery-edged tail.

Like the Pigmy Possum and the Sugar Glider, the Feathertail at night seeks out flowering trees, where it licks nectar from the open cup-shaped flowers of eucalypts, and eagerly snatches up any insects it finds.

The Feathertail Glider inhabits eucalypt forests and woodlands of eastern Australia, from south-eastern parts of South Australia to north-eastern Queensland, extending inland through woodland country on the western slopes of the Great Divide. It derives its common name from the feather-like fringing to either side of its long prehensile tail, an adaptation which presumably assists in controlling the direction of its gliding flights.

Compared with the lengthy glides of the larger species, the Feathertail seems mostly to make slightly prolonged leaps. When short jumps are compared with those of the Pigmy Possum, which has no hesitation in leaping about in the treetops, the gliding membrane does not seem to make a great deal of difference. But undoubtedly leaps from greater heights would give greater air speeds and a gliding descent. These animals are so small that the opportunity to see one leap from a high tree does not often occur.

Feathertails hide during daylight hours in a globular nest of gum leaves and bark tucked into a hollow or crevice of a tree. Often these nests contain a small group, sometimes a female and her family of up to four young, which is the maximum number that can be accommodated in her tiny pouch.

The Feathertail is the only Australian member of the genus *Acrobates*. Like the Pigmy Possum, it is more like the primitive ancestral type of possums, insect eaters with many pointed teeth. The Honey Possum, by comparison, has very degenerate and virtually useless teeth resulting from an almost purely nectar diet over the ages. The large gliders on the other hand eat also fruits and foliage and have herbivore dentition.

The tiny Feathertails, like the Honey and Pigmy Possums, are not often seen unless one of them happens to meet with some misfortune, such as being caught and brought in by a cat, or being found during tree-felling.

Spotted Cuscus

Phalanger maculatus

AUSTRALIA'S two species of cuscuses are rather foreign to the general Australian wildlife scene, having invaded this continent from the islands to the north. Having arrived here, they have never established more than a precarious hold upon a small part of Cape York.

The Spotted Cuscus is a large animal, with a head-and-body length of more than 60 cm. In general appearance it is more monkey than possum-like. The long strong prehensile tail, devoid of fur on the terminal half, has rasp-like scales right along its undersurface, and is commonly carried rolled into a tight coil. The ears are so small that in frontal view the animal appears to have almost no protruding external ears. In colour it is variable, rusty brown to grey, spotted or plain, sometimes completely white.

The second species of cuscus inhabiting Australia is the Grey Cuscus, which resembles the Spotted Cuscus but has a dark strip beginning between the ears and extending right down the back to the rump.

These cuscuses are common on the island to the south of Australia, from the Solomon Islands to the Celebes. In a past age, perhaps when lower sea levels exposed the bottom of the shallow seas of Torres Strait, the cuscuses, like many other animals, extended their range of distribution southwards to Cape York. In Australia they are now restricted to the rainforests and gallery forests of the eastern coast and ranges of Cape York, from the McIlwraith Range northwards.

The cuscuses are slow-moving inhabitants of the treetops, feeding at night upon native fruits and leaves, but also eating any small animal or birds which they may happen to capture. During the day they hide in dense foliage, such as the big clumps of ferns, epiphytic orchids, vine tangles, and similar places in the upper canopy of the forest.

These are of course marsupials, the young being carried at first in the pouch and, later, on the female's back. The cuscuses are members of the large family Phalangeridae, in which is included also the large brush-tailed possums, and the Scaly-tailed Possum. This latter animal, which is found only in rocky areas of the central Kimberley region of Western Australia, is in general appearance between the cuscuses and the brush-tailed possums, but its unusual tail is entirely covered with rasp-like scales instead of fur.

Distinguishing features of all the animals of this family are the very hand-like hind feet unlike the long hind feet of the kangaroos, the forward-opening pouches unlike the backward-opening pouches of the koala and wombats, and the chisel-like incisors of the herbivores rather than the pointed teeth of the marsupial carnivores.

South-west Pigmy Possum

Cercartetus concinnus

AMONG the smallest, most delightful and dainty of the possums, the South-west Pigmy Possum is one of a family of six species, all superficially mouse-like in size and general appearance. This family, which contains all the very small possums and gliders (family Burramyidae) is divided into three groups or genera. The family name comes from one of its genera, *Burramys*, which contains but a single species, a 'living fossil', the Mountain Pigmy Possum *Burramys parvus*.

Although at first glance looking mouse-like, the pigmy possums differ in that being insectivorous marsupials, they have many small pointed upper incisor teeth where ordinary mice have a pair of flat rodent chisel teeth. The females carry their young in a pouch. The prehensile nature of the tail is obvious with living specimens. It is able to curl around twigs as effectively as another hand. Their hind feet too are of the typical possum hand-like design, with opposable thumb, evolved for grasping when climbing.

The four pigmy possums of the genus *Cercartetus* are all very much alike, but fortunately there are few regions where more than a single species occurs, thereby making identification easier.

These little marsupials are solitary and nocturnal in habits, living in foliage of trees and shrubs in eucalypt forests and woodlands. By day they hide in small hollows of trees, under loose slabs of bark, in abandoned bird nests, in the dense dry 'beards' of grass-trees, and any other convenient places of concealment. At night they hunt for insects and feed avidly on the nectar of flowering trees. Not having the extremely elongated snout and long tongue of the Honey Possum, they are more or less restricted to flowers with an open nectary, like the cup-flowers of eucalypts. Moving through the foliage pigmy possums are surprisingly agile, leaping and clinging among the thinnest twigs and leaves, for they are so small that a single leaf will easily carry their weight. Although not possessing gliding membranes, the pigmy possums make leaps of perhaps a metre, on a downwards arc, to land clinging with outstretched limbs and clinging tail.

While sleeping in their nesting places pigmy possums are able to lower their body temperature, rather like a hibernating animal, becoming cool and torpid. Those found like this are generally rolled in a tight ball, with big soft ears folded down over their eyes, tail rolled. Only after some considerable time and seemingly with great difficulty are they able to begin to move, shaking and unsteady, befuddled by their premature awakening.

The South-west Pigmy Possum, or Mundarda (its Aboriginal name) occurs in two separate populations. One inhabits south-western Australia, the other the south-east of South Australia and across the borders into western Victoria and south-western New South Wales.

The Eastern Pigmy Possum, *Cercartetus nanus*, is very similar, and inhabits the forests and woodlands of coastal eastern and south-eastern Australia, and Tasmania. This island has a second species, the Tasmanian or Little Pigmy Possum, *C. lepidus*, which is noticeably smaller. It is found also on Kangaroo Island. The fourth species is the Long-tailed Pigmy Possum, *C. caudatus*, found in north-eastern Queensland and New Guinea. The remaining genus of the family has but a single species, *Acrobates pygmaeus*, the Feathertail or Pigmy Glider.

Koala

Phascolarctos cinereus

OF ALL the Australian herbivorous marsupials the Koala is probably the most highly specialised. It is entirely a creature of the treetops. The leaves of only a few species of smooth-barked trees are acceptable as food. No single species of food tree is adequate throughout the year, because at times the new leaves contain poisonous prussic acid, so the Koalas must at certain seasons change the species of their food trees.

About one kilogram of food is consumed each day, selected from Forest Red Gums, Sydney Blue Gums, Bush Box, Peppermint, Mahogany, Manna Gum, Messmate, or Swamp Gum.

The Koala is so distinctive that the single species is placed in a family of its own. There are no close relatives but there is a greater affinity with wombats than with any other marsupial group.

Although both the Koala and the possums are tree dwellers, and the wombats are terrestrial animals, studies of special features of anatomy, blood proteins and chromasomes indicate a closer relationship with the wombats than with the possums. Obvious outward similarities are the pouches which open towards the rear, and the extremely short tails.

The Koala would appear to be disadvantaged both by the lack of any useful tail for climbing, and the downwards facing pouch. For the burrowing wombats, the tail is no loss, and the rearwards opening pouch is an advantage. So it is possible that the Koala in some ancestral times was a ground-dwelling marsupial.

Apart from its dietary specialisation, the Koala is specialised for climbing, with long arms, and powerful claws with a vice-like grip.

The Koala gives birth usually to a single young, which is carried on the female's back after the usual period of time in the pouch. Koalas are therefore very slow breeders, probably averaging one young per year, which no doubt has contributed to their past decline in numbers in most regions.

The Koala inhabits the forests of eastern and south-eastern Australia, and fossil remains show that it previously occurred in south-western Australia.

Hairy-nosed Wombat

Lasiorhinus latifrons

AUSTRALIA has four species of wombats, which are placed by zoologists into two main groups, the plains wombats which have the lips of their snouts covered with fine hair, and the forest wombats which have bare snouts.

The forest wombats (genus *Vombatus*) are the best known, for their distribution coincides with the regions of greatest human population. Inhabiting ranges and forests along the east coast and in Tasmania, they are probably safe as wild creatures of viable populations for the foreseeable future. This forest wombat group includes the Common Wombat, the Tasmanian Wombat and the Islands Wombat of the Bass Strait islands.

The plains wombats (genus *Lasiorhinus*), inhabiting open grasslands and scrubby plains, have had their environment greatly changed through pastoral activities. Their habitats have not offered the same degree of security that the rough ranges and tall timber country have given the forest wombats. Consequently two species, the Queensland Hairy-nosed Wombat and the Moonie River Wombat of south-eastern Queensland, are now very rare. The third species of hairy-nosed wombat is still plentiful only in the most remote part of its range, which is the desert-like Nullabor Plain.

The plains or Hairy-nosed Wombat is distinguished not only by its silky-furred snout, but also by its much longer, more pointed ears. Fully grown, it is around a metre in length, some thirty centimetres in height, and weighs about thirty kilograms. Though short, its legs are powerful, with strong claws well adapted for digging.

The fur of the Hairy-nosed Wombat is much softer, more silky, than that of the common forest wombats, and of a silvery-grey colour.

Detailed studies have been made of the ecology of the Hairy-nosed Wombat, in an effort to protect both the animal and its environment.

Extensive burrows are constructed, rather like a rabbit warren, but on far larger scale. On the Nullarbor Plain, the burrows penetrate beneath the hard limestone cap-stone layer. In very hot weather the wombats will rarely be seen above ground during the day, conserving water and energy in the cool humid burrows. They are said to be able to go without water for three or four months in this way. But during mild sunny weather they may be seen lying lazily on the mounds of soil at their burrow entrances. Wombats, being marsupials, have pouches but, unlike the typical kangaroo-pattern of pouch, that of the wombat opens rearwards, perhaps because it is less likely to fill with sand during digging.

Quenda

Isoodon obesulus

AUSTRALIAN marsupials for the most part fall into either one or the other of two major groupings, being either carnivorous or herbivorous marsupials. An important exception, however, is the bandicoot family, the Peramelidae. They are therefore a third main branch of the marsupials, one which is intermediate between the others, for the bandicoots on the whole are mixed feeders, eating both plant and animal food. But within the bandicoot family there are considerable differences also, some genera being entirely insectivorous, others are entirely vegetarian.

The bandicoots, in their evolutionary position part-way from insectivore to herbivore, retain features of the one group and have acquired some of the features of the other.

Bandicoots may be recognised and distinguished from all other marsupials by their feet and their teeth. Because they are, unlike most of the carnivores, purely terrestrial, the structure of their hind feet is kangaroo-like, very elongated, and with combined second and third toes. These syndactyl toes, as they are called, look like a single small toe, but have two claws, which are used rather as a comb in grooming the fur.

But in contrast to these typically kangaroo-like, herbivore type hind feet, the bandicoots retain teeth more like those of the marsupial carnivores. Where the kangaroos, from the smallest potoroos upwards, have two broad chisel-like incisors at the front of the lower jaw, no bandicoot has less than three pairs of very small incisors between the two larger pointed canine teeth.

The bandicoot family is subdivided into two main groups, one containing the long-eared, silky-furred and rare rabbit-eared bandicoots, and the other the ordinary bandicoot and the pig-footed bandicoots.

The ordinary bandicoots in turn are divided into two groups, the long-nosed bandicoots, and the short-nosed. Both have sharply pointed snouts, and even the short-nosed bandicoots, compared with other animals, look long-nosed, until compared with a true long-nosed bandicoot.

The Quenda is also known as the Southern Short-nosed Bandicoot. It is a rabbit-sized animal, with a distribution along the east coast of Australia from Cape York to Adelaide, and also occurs in south-western Australia and Tasmania.

This nocturnal insectivorous bandicoot inhabits forests and other places where there is dense undergrowth. Its presence is often revealed by the little conical pits where it has been digging for beetles and various insect larvae. A pile of sticks and leaf litter forms a rough nest. There is often no obvious entrance, the bandicoot just pushing under the debris.

Brindled Bandicoot

Isoodon macrourus

MOST BANDICOOTS have harsh, almost spiny, fur, made up of long outer guard hairs beneath which is a softer under-fur. The fur of the Brindled Bandicoot is brown finely speckled with black, the brown being lighter and more golden on this bandicoot, as typical of specimens from the Kimberley and Arnhem Land.

The Brindled Bandicoot inhabits dense undergrowth of woodlands and forests of north-eastern New South Wales, eastern Queensland, and from Arnhem Land to the Kimberleys of Western Australia.

The common name given to this animal, 'bandicoot', is not of Australian origin, but was earlier used for a kind of large Indian rat.

It is now thought that there are probably only three species of short-nosed bandicoots. Older reference books list a considerably greater number, but these are probably only local variations of the three species.

One species, the Southern Short-nosed Bandicoot or Quenda, covers southern Australia from east to west. A second, the Golden Bandicoot, *Isoodon auratus*, ranges across central Australia from northern parts of South Australia through the Northern Territory to north-western parts of Western Australia. On Barrow Island, off the north-west coast, the isolated population of this bandicoot is sufficiently different to have been listed in the past as a separate species.

The third of the three short-nosed bandicoots, the Brindled Bandicoot, also has local colour variations as previously mentioned.

Bandicoots are one of the few kinds of native mammals to have remained abundant close to the major cities of Australia. The Brindled Bandicoot is still common near Darwin. The Southern Short-nosed Bandicoot, locally known as the Quenda, is the most common native mammal in the Darling Ranges near Perth. In the suburbs of Sydney it is the Common Long-nosed Bandicoot (*Perameles nasuta*) which is abundant.

Short-nosed bandicoots, being easily found, have been the subject of close study. They are strongly territorial, and aggressive towards other individuals of the same sex. The territories of the males are larger, and usually overlap the territories of several females. In the case of the Southern Short-nosed Bandicoot in Tasmania there may be two or three litters of three to eight young each breeding season, which lasts from May to February. In New South Wales and Queensland breeding is continuous throughout the year.

There are in the bandicoot family two most unusual forms which unfortunately have not remained common.

The Rabbit-eared Bandicoots or Dalgytes are probably one of the most graceful and attractive of native mammals, with soft silky fur, a flowing banner-like black-and-white tail, and long rabbit ears. They are largely desert dwellers, and now may be found only in remote central Australian and far northern areas.

So rare is the strange Pig-footed Bandicoot that no present-day biologist has seen one. The last record was from the Pitjanjarra tribesmen of the Musgrave Ranges, Central Australia, and even there it has not been seen since about 1926.

Platypus

Ornithorhynchus anatinus

THE PLATYPUS, probably the strangest of all Australian animals, appears to be a peculiar mixture of bird and mammal. But such unusual features as the duck bill and webbed feet are adaptations to the amphibious way of life, while the egg-laying rather than live birth is a characteristic retained from its reptile ancestors. In the water its webbed feet expand into broad paddles, though the hind feet usually trail behind, giving, together with the tail, stability and directional control. When it dives the ears and eyes are closed. The prey is found and identified by touch, using the sensitive bill.

The Platypus is often described as 'duck-billed', but this is a very superficial comparison. The bill or beak is not hard and horny like that of birds but has a framework of bone covered with soft moist skin naked of any fur. The bill increases in width towards its rounded tip, where a pair of nostrils is situated. The rest of the head is furred and merges into the body without any clearly defined neck. The Platypus has no external ears and the eyes are small. Both the ear apertures and the eyes are set into a deep groove in the skin, the edges of which close together when the animal is under water. This probably makes the Platypus both blind and deaf once it submerges.

But it is in the underwater environment that the marvellous bill takes over, guiding the Platypus along waterways and creekbeds, around boulders, under and over logs, and among reeds, finding and identifying the prey. The skin of the bill is a mass of extremely sensitive nerve endings, by which the Platypus feels its way under water, snapping up its prey and coming then to the surface to feed.

Although generally thought of as aquatic animals, Platypuses actually spend only about two hours of every twenty-four in the water, feeding. The pattern is generally to emerge from the burrow in the early morning and again in late afternoon. In the water the front feet are the paddles. While floating at the surface, the Platypus is always alert. Both sight and hearing are then very acute.

The Platypus when feeding stays submerged for periods of one to five minutes. When floating on the surface it can easily be overlooked because it lies very low in the water and looks just like any piece of dark waterlogged wood.

The burrow excavated by the Platypus may be from 4.5 metres up to 9 metres long, but even as long as 27 metres. The entrance is usually well above water level. The tunnel is plugged at intervals with soil while the Platypus is inside, perhaps to deter predators, or to maintain a warm humid atmosphere.

In a widened chamber a nest of grass and gum leaves is constructed and two soft-shelled eggs laid. The Platypus, like the Spiny Anteater, has no nipples, the milk being sucked up by the young as it exudes from the pore-like ducts of the mammary glands.

Since the time of the appearance of the earliest reptile-like mammals, the evolution of these furred animals has followed three almost parallel paths of development. Probably the earliest were the monotremes, followed by the marsupials and most recently the placental or eutherian mammals, but all three forms still exist today.

The monotremes, most primitive in that they were most like their reptile ancestors, seem to have been the least successful. Except for those relic species, the Platypus and Spiny Anteater, they have been displaced by the marsupials.

Spiny Anteater

Tachyglossus aculeatus

ONE OF the world's very few species still surviving of the monotreme order of mammals, the Spiny Anteater or Echidna shares with the Platypus the distinction of being a primitive link between the reptiles and the other more advanced mammals. Like the Platypus the Spiny Anteater lays eggs, has a reproductive system like that of reptiles, and resembles them also in skeletal features and in its rather variable body temperature. Although very widespread, from rainforest to semi-desert, it is an inconspicuous creature in its natural habitat.

In apparent contradiction to its primitive make-up, the Spiny Anteater has remained one of the most successful of Australian mammals and has survived where many species of marsupials and higher mammals have been decimated by changes of environment, predation or competition. The reasons for this success are probably twofold: the protection of its spiny armour and its extremely specialised way of life.

The Spiny Anteater feeds upon ants, for which very few other Australian animals have any liking. The Marsupial Anteater or Numbat does not compete, for it seeks only termites.

The Spiny Anteater's long tube-like muzzle, like the 'beak' of the Platypus, is an adaptation to a way of feeding. Its generic name, *Tachyglossus*, means 'quick-tongued'. The small mouth opens just below the tip of the muzzle and the incredible tongue can extend about 30 cm. It is sticky and flickers quickly in and out to pick up the meal of ants.

Confident of the protection given by the coat of spines, the anteater is out during the daylight hours and is most active during early afternoon. This is generally the warmest time of day, when ants tend to be most active. There would be few ants to be found at night, so the anteater is by necessity a diurnal animal. Being, as typical of reptile-like monotremes, of low-slung build, and thick-set for the strength to rip into ant nests and tear away debris, the anteater is hardly suited to making a fleet-footed escape from its enemies. Deprived also of the opportunity to feed after dark, it has evolved an admirable solution: the spiny armour which enables the anteater to go about its daily business almost oblivious of the predators of the bush.

If disturbed, the Spiny Anteater's instant reaction is to burrow straight downwards. Once half-buried, with the massive claws of its incredibly strong limbs hooked under roots or rocks, it is almost impossible to lift up or turn over. But if the anteater does happen to be taken by surprise on a hard smooth surface where it cannot burrow or grip, it curls into a ball, rolling to the inside its soft undersurface and face, and presenting outwards a radiating sphere of long needle-pointed spines.

Most of the anteater's head and undersurface is covered with fur, with small spines on the back of the head and very long spines on the back and sides. The Tasmanian Spiny Anteater has much longer fur, almost hiding the spines.

The female anteater lays a single egg, which she puts into a temporary pouch on the underside of her body. It appears that she lies on her back and curves her body so that the egg when laid rolls into this pouch. After seven to ten days it hatches, and the tiny hairless baby anteater lives in the pouch for six to eight weeks, until its sprouting spines make it necessary for the female anteater to put it in a safe hiding place until it can fully fend for itself.

Dingo

Canis familiaris

AUSTRALIA'S single species of wild dog, the Dingo, as well as being the sole representative of the canines (family Canidae) is also the only species of indigenous land carnivore. The true cats are not represented among Australia's native fauna, while the seals, also carnivores, merely touch upon our coasts.

In the past, all the predatory niches of the environment have been occupied by marsupials of carnivorous way of life. They are not related to the ordinary carnivores of other continents. These marsupials had evolved a variety of species for this continent to parallel those of other lands, e.g. the dog-like Thylacine or Tasmanian Tiger, the hyaena-like Tasmanian Devil, the cat-like native-cats and, in past ages, a lion-like marsupial carnivore, the Thylacoleo.

But the Dingo can only just be considered one of Australia's native mammals, for it is a comparatively recent arrival from islands or continents to the north, probably being brought here as a domesticated dog by the Aboriginals who came via the islands. Since that time many thousands of years ago the Dingo has spread throughout the continent, although never reaching or being established in Tasmania.

The Dingo arrived as a predator higher up the evolutionary scale than the comparatively primitive marsupial carnivores, which had always been isolated on this island continent and thereby sheltered from direct competition from the carnivores of the rest of the world. A more efficient hunter, the cunning Dingo took the prey of the marsupial carnivores, and was probably responsible for the disappearance of the wolf-like *Thylacine* from the Australian mainland. This may also account for the extinction of the Tasmanian Devil except in Tasmania. These marsupials have gone in comparatively recent times. The Thylacine is shown in old Aboriginal cave paintings in the Northern Territory, while mummified remains found in a cave of the Nullarbor still had the characteristically striped fur. The Tasmanian Devil's remains have been found in many caves and other excavated sites.

Thus it would appear that we have the Dingo, which is not sufficiently different from ordinary dogs to be a separate species, at the expense of several truly unique species of marsupial carnivores, which but for the Dingo may have been reasonably common on mainland Australia today. In addition, the Dingo may have placed a heavier pressure of predation upon many small wallabies and the like. This effect which an introduced animal may have upon native animals is an all too familiar tale of tragedy in Australia, for European man has been far more damaging with his introductions, the cat, fox, rabbit, goat, rat, buffalo, sheep, cattle, camel, donkey and so on. All are having a devastating effect upon native animals or vegetation in one part or another of Australia, whether it be the killing of birds by cats or the overgrazing of arid lands by sheep and cattle.

The close relationship between the Dingo and ordinary domesticated dogs is shown by their interbreeding and the difficulty of distinguishing many Dingoes from ordinary dogs. A number of minor features are a guide—the ears remain erect, the tail is brushy, the canine teeth are on average larger, it yelps or howls but does not bark, and there are behaviour differences. The colour of wild Dingoes is variable and not a reliable guide, though tawny yellow is common.

Spinifex Hopping-mouse

Notomys alexis

IN APPEARANCE these small mammals are like miniature kangaroos, to which they are not even distantly related, being rodents not marsupials. The hopping-mice, known also as kangaroo-mice, can easily be recognised as rodents by their chisel-like paired incisor teeth and the absence of any pouch on the female. They are most likely to be confused with the *Antechinomys* marsupial-mice, which have a superficially similar appearance, with long tail and more or less hopping or leaping action. But being marsupial insectivores, the *Antechinomys* have needle-pointed teeth.

The Spinifex Hopping-mouse is widespread through the interior of Australia, from western Queensland through Central Australia and northern South Australia to central Western Australia. It prefers desert dunes and sandy spinifex grasslands.

If pursued the hopping-mice are extremely fast for such small creatures, dodging between clumps of spinifex with abrupt changes of direction. Spinifex is a tough spiny grass which forms dense clumps with areas of bare sand between. The extremely long tails of the hopping-mice are brush-tipped. They probably act as rudders trailing behind, helping the mice to make sudden turns at speed, when they are touching the ground only momentarily at each leap.

The biology of this and several other kangaroo-mice species has been the subject of intensive study. They are able to live in the arid deserts of the interior because they are nocturnal, avoiding the intense heat of the day in burrows about a metre deep. Although these mice will drink water if it is available, they are able to thrive, even gaining weight, on a diet of dry grass seeds, without drinking any water at all. They achieve this through their remarkable ability to give out a minimal amount of water in urine, which is more concentrated than for any other known rodent.

The burrows of the hopping-mice are complex and cunningly constructed. Burrows were found by the author in spinifex-covered sand dune country of arid north-western Australia. Trails of tiny footprints in the sand converged upon the entrance holes, which were usually situated in the clear spaces between the spinifex clumps. But hidden under or close to the spinifex were numerous escape exits, as many as nine for a single burrow complex. If any predator tries to dig down to the mice, it will certainly find the burrow deserted and the mice escaped by one of their many escape exits.

Upon excavating the tunnel it was found that the entrance burrow descended vertically, reaching a depth of about a metre where there was a plug of soft soil, then the tunnel turned sharply to run horizontally. About two-thirds of a metre along this main horizontal tunnel was a nest chamber approximately 10 cm wide, 30 cm long and 7.5 cm high. From the far side of the chamber were two escape tunnels, sloping upwards at first, then turning vertically towards the surface, where each was blocked with a plug of sand.

In the centre of the nest chamber was found a cup-shaped grass nest, 7.5 cm in diameter, resembling the nest of a bird, but set into a hollow in the floor of the chamber. It contained three almost newborn hopping-mice, still blind and completely naked of fur. Their mother was found in one of the escape tunnels, digging towards the surface through the soft sand plug, packing sand behind her as she went as a barrier between herself and the intruders.

Western Chestnut Native-mouse

Pseudomys nanus

THIS VERY SMALL native mouse which at first appears identical to an ordinary house mouse is a member of a group of native rodents which are found only in Australia, endemic to Australia, and have been following their own evolutionary paths in this isolated continent for a very long time. They are referred to, therefore, as the 'old endemics', a term of descriptive convenience.

Within this group are many of the most interesting of the Australian rodents. During their long period of isolation the various rodents of this group have found their place within the overall ecology. Each has become specialised for its own particular place in the environment. As a result we now have groups of animals that in their appearance or way of life are not only unique to Australia but are also unique as unusual kinds of rodents.

The old endemics include the hopping-mice, which are like miniature kangaroos but live in deep and cunningly complex burrows and are highly adapted to desert survival. There are also the rabbit-like stick-nest rats, which build 'houses' of sticks and shelter within them from predators and from the desert heat. Rock-rats, inhabitants of stony areas of central and northern Australia, have peculiarly thickened long tails, rather like the tails of some of the marsupial mice. Their tails are ridged in a ring-like pattern and are very delicate, the flesh stripping away easily, a parallel to the lizard trick of shedding its tail when caught.

The tree-rats too are old endemics. One of these, the Golden-backed Tree-rat, is one of the most beautiful of the native rodents. This species, together with the Black-footed Tree-rat and the Brush-tailed Tree-rat, is still quite common in northern Australia, but the White-footed Tree-rat of eastern Australia has not been seen this century. These quite large, agile animals apparently feed on the ground, but take refuge in trees.

A relic species, probably edging towards extinction, is the Broad-toothed Rat, which, although once much more widely distributed, now survives only in a few colonies in very cold alpine areas of Tasmania, New South Wales and Victoria, where it inhabits boggy sedges, button-grass and sphagnum moss areas.

Less interesting in appearance and habits, but unique to Australia as part of the old endemic group, is the large genus *Pseudomys*, the native mice which include the Western Chestnut Native-mouse. Relatively little is known about most of these rather insignificant little rodents, of which there are about seventeen species spread across much of Australia. The Western Chestnut Native-mouse occurs in the north-west and in the Northern Territory, in rocky country. This particular individual was found near the Drysdale River, in the north Kimberley region. It is about the size of a house mouse, with glossy fur, and a tail about the same length as the combined head-and-body length.

Common Rock-rat

Zyzomys argurus

AUSTRALIA has three species of native rats which are found only in rocky outcrops of Central Australian and far northern regions. This species, the Common Rock-rat, is native to the Northern Territory, north Queensland, and the Pilbara and Kimberley areas of Western Australia. It is somewhat smaller than the ordinary introduced rat, but substantially larger than the house mouse. The fur is yellowish-brown on the upper surfaces, paler on the underparts of the body.

A rather deceptive feature is the tail which on young animals is an ordinary slender rat tail, with rather sparse hairs longer towards the tip. Clearly visible beneath these hairs the tail is ridged to give a ringed appearance. On adult rock-rats, however, the tail becomes greatly thickened, rather like the tails of some marsupial-mice. If roughly handled, caught or restrained by the tail, the flesh very easily strips away from the vertebrae. If the animal escapes its attacker and survives, this section of vertebrae is soon shed, giving a permanently shortened tail, mammals not having the ability of reptiles to regrow lost tails.

Apart from the Common Rock-rat, there are two other species, which also have the thick tails. The Large or Woodwards Rock-rat *Zyzomys woodwardi* is larger, solidly built rather like a guinea-pig in body shape with harsh grizzled fur. The MacDonnell Range Rock-rat *Zyzomys pedunculatus* is around the size of an ordinary rat, has soft brown fur, and the tail is brown above, pale below. These two species were formerly placed in a separate genus, *Laomys*, and called thick-tailed rats.

The rock-rats are members of the old endemic group. The females share with others of the group the curious habit of dragging their young about, even as these grow quite large, attached tenaciously to their teats.

In any discussion of Australian native rats, the aquatic species, the Water-rat and the False Swamp-rat, are worthy of mention, being species which during their very long time on this continent have found a 'niche' in the ecology and have successfully specialised for this way of life. The Water-rat shares the freshwater rivers and lakes with the Platypus, but is more widely distributed, being found in most parts of Australia where there are rivers or lakes, and on some ocean beaches. There are some distinctively different coloured regional variants.

The Water-rat is large, up to 60 cm in length. It has webbed hind feet, very dense soft fur, and a white-tipped tail. It is an entirely carnivorous animal, living on freshwater crayfish, fish, lizards, even occasional water birds, and shellfish such as mussels. It has learned to open the mussels without damaging the shells. It places the mussels out of the water where the heat of the sun causes them to open.

The False Swamp-rat is a smaller aquatic species found in north Queensland and the Northern Territory. It appears to be quite rare. Although generally similar to the water-rat, it is smaller, has a shorter tail, and the hind feet are not webbed.

To give an overall picture of the native rodents of Australia, mention must be made of the mosaic-tailed rats, which include the Great White-tailed Rat *Uromys caudimaculatus* of north-east Queensland rainforests, and the three species of *Melomys*. All have tails with a random mosaic pattern of scales instead of the usual rat arrangement of scales in rings around the tail.

Tunneys Rat

Rattus tunnei

THIS MEMBER of the Australian bush-rat group is an inhabitant of the northern half of Australia, north of a line from Rockhampton through Alice Springs to the mid-west coast of Western Australia. It occurs in the Kimberleys, the 'top end' of the Northern Territory, and on Melville Island. Within this region this species, which is known also as the Paler Field-rat, favours river flats, heathlands, grassy woodlands, and sandy coastal areas.

Superficially this native rat looks not greatly different from others of its kind. Compared with the Southern Bush-rat, its fur is sandy brown rather than grey, and long, giving the animal a rounded fluffy appearance and making the tail, already distinctively short, look even shorter. The belly fur is entirely whitish without the usual greyish base. A relatively stoutly built rat, it has a rather broad skull. This species was originally described from a specimen from the Mary River in the Northern Territory, in 1904. A sub-species from the Alice Springs area is described as being smaller, more brightly coloured and with thicker fur, while another sub-species from coastal Western Australia is of similar colour to the north-coastal animal but slightly smaller.

Like the bush-rats this species still bears the generic name *Rattus*, indicating that it is not greatly different from the rats of other continents, having been in Australia for a period of time insufficient to evolve really substantial differences of anatomy. It has been in Australia for perhaps tens of thousands of years.

There are other groups of native rats and mice known as the 'old endemics', meaning that they have been in Australia for a much longer time, and show a greater degree of differentiation from the rats of other parts of the world.

The history of the Australian fauna is a sequence of invasions. Probably only the monotremes and marsupials are the true 'original' Australian mammal fauna. The widely accepted theory of continental drift describes Australia as breaking away from Antarctica and other land masses after the evolution of the earliest mammals. During the following 50 million years or thereabouts the monotremes and marsupials were probably Australia's only mammal inhabitants, until comparatively recent times, when the sea gap northwards to Asia narrowed, and sea levels dropped during the succession of ice ages.

The series of mammal invaders, first the ancestors of our 'old endemics', then other rodents, included some 50 thousand years ago the first of the Aboriginals, and the Dingo. The process has accelerated recently with the introduction of rabbits, foxes, cats, sheep, cattle and many other additions to the mammal population, resulting in the loss of some of the creatures whose history has been linked with that of the continent of Australia for a tremendous length of time.

Southern Bush-rat

Rattus fuscipes

THE AUSTRALIAN bush-rats are true native animals. Although they belong to the same genus as the repulsive introduced rats, these native rats have been inhabitants of this continent for many thousands of years.

The native rats are clean and quite attractive little animals, with soft glossy fur. Being exclusively bush dwelling, never found in the cities, or in garbage, sewers or other man-made sources of disease, these mammals are as clean as any other bush creatures.

The bush-rats are not marsupials but placental mammals. They have no pouch, but give birth to comparatively large and well-formed young as do all other placental mammals. The term placental indicates that the young, by means of a placental attachment and umbilical cord, are able to be retained in the womb until well developed, whereas the marsupial, without this advanced placenta, gives birth to young at an almost embryonic stage and provides a pouch for completion of development.

The identification of Australia's many species of native rats and mice is much more difficult than for the marsupials as there are but small external differences between many species. In some instances, positive identification can only be made by expert examination of details of teeth and skulls.

External characteristics used in identification include the relative lengths of tail compared with combined head-and-body lengths, and the number and distribution of nipples along the undersurface of the body.

The bush-rat group contains about seven species, all being in the genus *Rattus*. Of these, the Southern Bush-rat *R. fuscipes*, now includes what were three separate species which, upon further examination, have been recognised as being merely geographical variants of this one species. Those three former species were the Allied Rat *Rattus assimilis* of eastern coastal Australia, Grey's Rat *R. greyii* of South Australia, and the Western Swamp-rat *R. fuscipes* of south-western Australia.

Old reference books will be found to list these as separate species. Today all are forms of the species *R. fuscipes*, taking that name because that was the first named, in 1839. All forms are given the common name of Southern Bush-rat. This has been seen as being the most generally appropriate.

Other species of bush-rats which still remain as separate distinct species include the Eastern Swamp-rat *R. lutreolus*, the Dusky Field-rat *R. sordidus*, the Long-haired Rat *R. villosissimus*, the Mottle-tailed Cape York Rat *R. leucopus*, and Tunneys Rat *R. tunnei*. The Southern Bush-rat may be found in many habitats from rainforests to coastal heathlands, and in dense low vegetation around swamps. The presence of the Southern Bush-rat can be detected often by the tunnel-like passageways through the dense low vegetation. This rat breeds throughout the year. The usual three or four young are weaned at about twenty-five days. An average female will produce three or four litters in her life span of about one year.

INDEX